MODULAR MATHEMATICS

Module F: Mechanics 2

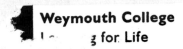

L. Bostock, B.Sc.

S. Chandler, B.Sc.

Stanley Thornes (Publishers) Ltd

First published in 1995 by Stanley Thornes (Publishers) Ltd,
Ellenborough House, Wellington Street, CHELTENHAM GL50 1YW

A catalogue record of this book is available from the British Library.

ISBN 0–7487–1774–9

Cover photograph: Martyn F. Chillmaid

Artwork by Mark Dunn
Typeset by Tech-Set, Gateshead, Tyne & Wear
Printed and bound in Great Britain at T J Press, Padstow, Cornwall

CONTENTS

Explanatory notes   v

Chapter 1 **Elastic Strings and Springs**   1
Extension of an elastic string. Hooke's Law. Modulus
of elasticity. Extension and compression of a spring.
Stretched strings and springs in equilibrium.
Mathematical modelling.

Chapter 2 **Motion in a Horizontal Circle**   16
Angular speed and velocity. Acceleration of a particle
moving on a curve. Motion in a circle with constant
speed. Relation between angular and linear speed.
Magnitude of radial acceleration. Forces producing
motion in a horizontal circle. Conical pendulum.
Banked tracks.

Chapter 3 **Work and Energy**   45
Work, kinetic energy, potential energy. Work–energy
principle. Conservation of mechanical energy. Work
done in stretching an elastic string. Elastic potential
energy. Work and energy problems.
Refining mathematical models.

Consolidation A   65

Chapter 4 **Motion in a Vertical Circle**   77
Motion in a circle with variable speed. Circular
motion in a vertical plane; restricted to a circular path
and not so restricted.

Chapter 5 **Collisions   Law of Restitution**   100
Momentum. Impulse. Relation between momentum
and impulse. Elastic impact. Coefficient of restitution.
Direct collision with a fixed object. Direct collision
between objects both free to move. Multiple impacts.

Chapter 6 **Centre of Mass**   121
Centre of mass. Centre of mass of compound laminas.
Centre of mass of a solid of revolution by integration.
Use of symmetry. Centre of mass of a semi-circular
lamina. Centre of mass of compound solids.

Chapter 7 **Equilibrium of Rigid Bodies**   143
Equilibrium of coplanar concurrent forces and of
coplanar parallel forces. General conditions for
equilibrium of a set of coplanar forces. Bodies
suspended in equilibrium. Bodies resting on an
inclined plane.

          Consolidation B   175

Chapter 8 **Variable Acceleration**   192
Motion in a straight line with a variable acceleration
that is a function of time. Acceleration as a function
of displacement; using $v\dfrac{\mathrm{d}v}{\mathrm{d}s}$ for acceleration. Velocity
as a function of displacement. Basic equation of, and
properties of, simple harmonic motion (SHM).
Associated circular motion.

Chapter 9 **Variable Force**   225
Use of $F = ma$ when acceleration is a function of
time or displacement. Work done by a variable force.
Impulse of a variable force. Forming, testing and
improving mathematical models. Newtons law of
universal gravitation. Simple pendulum. Seconds
pendulum. Forces producing SHM.

          Consolidation C   256

          Answers   267

          Index   279

EXPLANATORY NOTES

This book continues the mechanics section of modular mathematics courses for students aiming at post-GCSE academic qualifications.

Together with Module E this book covers the mechanics needed for A-level Mathematics syllabuses containing Pure Mathematics and Mechanics. Combined with other Modules, it also prepares students for courses such as AS Mechanics, A-level Applied Mathematics and A-level Further Mathematics. Some of the topics begun in Module E are developed further in this book; in all such cases, preliminary notes and an exercise are given for revision purposes.

All new work is fully explained together with worked examples to illustrate possible methods of solution; the exercises that follow contain a generous provision of questions, starting with straightforward ones, to help the reader develop confidence and proficiency before moving on to the next area of new work. At intervals there are Consolidation sections containing a summary of the work covered in the preceding chapters and an exercise comprising past and specimen examination questions. These questions are best used for revision and practice when the whole course has been completed and when confidence and sophistication of style have been developed.

In this book there are two broad approaches to mechanics:

1. Understanding and being able to use the basic laws and principles involved as they apply to theoretical situations.
2. Considering the use of simple models to solve real-life problems; the choice of a reasonable model, refining it and, where this is possible, testing it.

The accuracy to which answers are given depends on which of these skills is being applied. In a theoretical question it is reasonable to give answers corrected to three significant figures (we use 3sf to indicate that the third figure is *corrected*). On the other hand, since constructing a model already involves approximation, it is not usually appropriate to give answers corrected to more than two significant figures (2 sf). Angles are usually given to the nearest degree. The expected degree of accuracy is specified for some, but not all, of the questions in this book. In this way some guidance is given but the reader is also given the opportunity to decide on a sensible degree of accuracy.

We are grateful to the following examination boards for permission to reproduce questions from their past examination papers and specimen papers. Specimen questions are indicated by the suffix s and it should be noted that they have not been subjected to the rigorous checking and moderation procedure by the Boards that their examination questions undergo. (Any answers included have not been provided by the examining boards; they are the responsibility of the authors and may not necessarily constitute the only possible solutions.)

University of London Examinations and Assessment council (ULEAC)
Northern Examinations and Assessment Board (NEAB)
University of Cambridge Local Examinations Syndicate (UCLES)
The Associated Examining Board (AEB)
Welsh Joint Education Committee (WJEC)
University of Oxford Delegacy of Local Examinations (OUDLE)

L. Bostock
1995   S. Chandler

CHAPTER 1

ELASTIC STRINGS AND SPRINGS

ELASTIC STRINGS

When a string can be stretched by forces applied at its ends it is called an *elastic string*.

The *natural length* of an elastic string is its *unstretched length*.

Two forces, one at each end, must be applied to an elastic string in order to stretch it. You may argue that pulling *one* end will cause stretching if the other end is tied to a fixed object such as a wall, but remember that a force acts on the string at the point of attachment, e.g. where it is fastened to the wall.

Clearly the stretching forces must each act outwards; they must also be equal and opposite as otherwise the string would move in position and not just stretch.

An elastic string that has been stretched is *taut*; when it is in a straight line and is of natural length, it is described as 'just taut'.

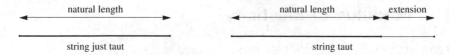

The difference between the stretched length and the natural length is called the *extension* and is often denoted by *x*.

We know that a taut string exerts an inward pull at each end and, by considering the equilibrium at one end of the string, we see that the tension is equal and opposite to the stretching force there.

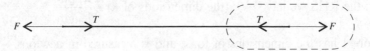

1

HOOKE'S LAW

In the seventeenth century a relationship between the extension of a stretched string and the tension at each end was discovered experimentally by Hooke.

The relationship, known as Hooke's Law, states that, up to a certain point,

the extension, x, in a stretched elastic string
is proportional to the tension, T, in the string

i.e. $T \propto x$ or $T = \dfrac{\lambda x}{a}$

where a is the natural length of the string

and λ is the *modulus of elasticity* of the string.

The Elastic Limit

As the extending forces applied to the string are steadily increased, there comes a time when a further increase suddenly produces an extension much greater than Hooke's Law would suggest. The string has become *overstretched* and will not return to its natural length when it is released; it has gone beyond its *elastic limit*. Subsequently its extension bears no relationship to the tension and, at this level of study, is no longer of any interest to us; in this book we deal only with strings that have not exceeded their elastic limit.

The Modulus of Elasticity

The form in which Hooke's Law is usually used is

$$T = \frac{\lambda x}{a}$$

Considering the dimensions on each side of this formula we see that

on the LHS we have   the dimensions of a force

on the RHS we have   (the dimensions of λ) $\times \dfrac{\text{length}}{\text{length}}$

Therefore λ has the dimensions of force and is measured in newtons.

Further, when the length of an elastic string is doubled, i.e. when $x = a$, then $T = \lambda$ showing that

**λ is equal to the tension in an elastic string
whose length is twice the natural length.**

SPRINGS

Hooke's Law applies to springs in a similar way as to elastic strings but there is one important difference – a spring can be compressed as well as stretched.

When stretched, i.e. when it is *in tension*, a spring behaves in exactly the same way as a stretched elastic string, i.e. equal and opposite *tensions* act *inwards* at the ends.

When a spring is compressed, the reduction in its length is called the *compression* and the forces in the spring are an *outward push*, called a *thrust*, at each end.

The spring is said to be *in compression* and it obeys Hooke's Law where T is the thrust and x is the compression.

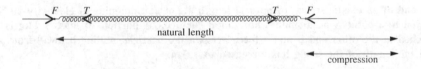

Examples 1a

1. **A light elastic string whose natural length is 0.8 m is stretched to a length of 1.1 m by a force of 12 N as shown. Find the modulus of elasticity of the string.**

$$\text{Tension} = \text{extending force} \quad \Rightarrow \quad T = 12$$

$$\text{Using Hooke's Law gives} \quad T = \lambda\left(\frac{0.3}{0.8}\right)$$

$$\therefore \quad \lambda = 32$$

The modulus of elasticity is 32 N.

2. **The natural length of a light spring is** a **and its modulus of elasticity is** λ. **Find, in terms of** a **and** λ, **the length of the spring when**

   (a) **the tension in the spring is** $\frac{1}{2}\lambda$

   (b) **the thrust in the spring is** $\frac{1}{2}\lambda$.

   When there is a tension in the spring it is extended and when there is a thrust it is compressed.

   (a)

   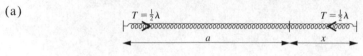

   Hooke's Law gives $\frac{1}{2}\lambda = \lambda \times \dfrac{x}{a}$ $\Rightarrow$ $x = \frac{1}{2}a$

   The extended length is $a + \frac{1}{2}a = \frac{3}{2}a$

   (b)

   $T = \frac{1}{2}\lambda$ $T = \frac{1}{2}\lambda$

   Again Hooke's Law gives $x = \frac{1}{2}a$ but this time it is a compression

   The compressed length is $a - \frac{1}{2}a = \frac{1}{2}a$

3. **One end of an elastic string, of natural length 1.2 m and modulus of elasticity 20 N, is fixed to a point A on a smooth horizontal surface. A particle of mass 1.2 kg is attached to the other end B and a force acts on the particle, pulling it away from A, until the length of the string has increased to 1.5 m.**

   (a) **Find the tension in the string.**

   (b) **If the force ceases to act on the particle, find the acceleration with which the particle begins to move.**

   (c) **State, with a reason, whether or not the particle continues to move with constant acceleration.**

   (a)

   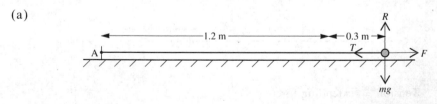

   The extension in the string is 0.3 m

   Hooke's Law gives $T = \dfrac{20 \times 0.3}{1.2} = 5$

   The tension in the stretched string is 5 N.

(b)

When the force is removed the only horizontal force acting on the particle is the tension in the string and this acts towards A.

Using Newton's Law $F = ma$

gives $5 = 1.2\,a \quad \Rightarrow \quad a = 4.17 \quad (3\text{ sf})$

The particle begins to move towards A with acceleration $4.17\,\text{ms}^{-2}$ (3 sf).

(c) As soon as the particle begins to move towards A the string gets shorter and the tension in it reduces, causing a reduction in the acceleration.

Therefore the particle does not move with constant acceleration.

EXERCISE 1a

1. An elastic string of natural length 1 m is stretched to a length of 1.3 m by a force of 3 N. Find its modulus of elasticity.

2. A light elastic string whose modulus of elasticity is 18 N is stretched from its natural length of 1.4 m to a length of 1.8 m. Find the tension in the string.

3. The length of a spring whose modulus of elasticity is 30 N is reduced by 0.4 m when compressed by forces of 20 N. What is the natural length of the spring?

4. A force of 8 N acts outward at each end of a light elastic string of natural length 1.6 m. Find the stretched length of the string if the modulus of elasticity is

   (a) 10 N   (b) 20 N   (c) 40 N.

5. The natural length of a light spring is 0.9 m and its modulus of elasticity is 20 N. Two forces act inwards, one at each end of the spring. Find the compressed length of the spring if the magnitude of each force is F newtons, where

   (a) $F = 10$ (b) $F = 14$ (c) $F = 4$.

6. A force F newtons acts inwards at each end of a light spring and produces a compression of 0.3 m. Find the value of F if the modulus of elasticity of the spring is 12 N and the natural length is

   (a) 0.5 m   (b) 1 m   (c) 2 m.

7. A light elastic string, with modulus of elasticity 22 N, is extended by 0.5 m. Find the natural length of the string if the tension in it is

   (a) 8 N   (b) 10 N   (c) 16 N.

8. One end of an elastic string of natural length 0.5 m is attached to a point A on a smooth horizontal surface. A particle P, of mass 0.4 kg is attached to the other end of the string. This particle is held on the table, at a distance 0.8 m from A, by a person exerting a horizontal force of 6 N.

    (a) Find the modulus of elasticity of the string.

    (b) The person releases the particle. Find its initial acceleration.

    (c) Find the tension in the string and the acceleration of P when the distance AP is,

        (i) 0.7 m   (ii) 0.6 m   (iii) 0.5 m.

9. An upper body exerciser consists of a spring of length 20 cm with a handle attached at each end. The modulus of elasticity of the spring is 600 N. A boy can stretch the spring to 25 cm in length. Find the force which he is then exerting on each handle.

10.

    Another model of the upper body exerciser, which uses springs of the same type as in Question 9, is adjustable by inserting extra springs between the handles. Ben uses two springs and can extend it by 13 cm. Tony uses three springs and can extend it by 9 cm. Who is the stronger?

EQUILIBRIUM PROBLEMS

The equilibrium of a particle under the action of coplanar forces was discussed in Module E; here is a summary of the results.

Two forces that keep a particle in equilibrium must be equal and opposite.

When **three forces** maintain equilibrium then

(a) lines drawn to represent the forces in magnitude and direction, form a triangle called a 'triangle of forces'

(b) they satisfy Lami's Theorem, i.e. the magnitude of each force is proportional to the sine of the angle between the other two forces.

When **any number of forces** acting on a particle are in equilibrium, the sum of the force components in any direction is zero. The forces are not necessarily concurrent and in this case the sum of the moments about any axis must also be zero.

Using these facts we can now look at some problems in which elastic strings or springs provide one or more of the forces that keep a particle in equilibrium.

EXAMPLES 1b

1. A light elastic string, of natural length 1 m and modulus of elasticity 35 N, is fixed at one end and a particle of mass 2 kg is attached to the other end. Find the length of the string when the particle hangs freely in equilibrium.

From the equilibrium of the particle,

$$T = 2g$$

Now using Hooke's Law we have

$$T = \frac{\lambda x}{a} \quad \Rightarrow \quad 2 \times 9.8 = 35 \times x$$

$$\therefore \qquad x = 0.56$$

Therefore the length of the string is 1.56 m.

2. The natural length of a light elastic string AB is 2.4 metres and its modulus of elasticity is $4g$ newtons. The ends A and B are attached to two points on the same level and 2.4 m apart, and a particle of mass n kg is attached to the midpoint C of the string. When the particle hangs in equilibrium, each half of the string is at $60°$ to the vertical. Find, corrected to 2 significant figures, the mass of the particle.

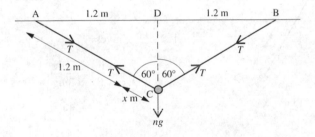

From symmetry the tensions in the two portions of the string are equal.

For the forces acting on the particle:

Resolving vertically

$$2T\cos 60° = ng \quad \Rightarrow \quad T = ng \qquad [1]$$

Now using Hooke's Law for either portion of the string,

$$T = \frac{\lambda x}{a} \quad \Rightarrow \quad T = 4g \times \frac{x}{1.2} \qquad [2]$$

From [1] and [2],

$$ng = \frac{4gx}{1.2} \quad \Rightarrow \quad x = 0.3n$$

In $\triangle ACD$, $\qquad AD = AC \sin 60°$

$\therefore \qquad\qquad AC = \dfrac{1.2}{\sin 60°} = 1.385\ldots$

Also AC is the stretched length of one half of the string

$\therefore \qquad\qquad AC = 1.2 + x$

$\Rightarrow \qquad 1.385\ldots = 1.2 + 0.3n \qquad \Rightarrow \qquad n = 0.618\ldots$

The mass of the particle is $0.619\,\text{kg}$ (3 sf).

3. **A light elastic string of natural length $2a$ is fixed at one end A and carries a particle of mass $3m$ at the other end. When the particle is hanging freely a horizontal force $3mg$ is applied to it. When the particle is in equilibrium the string is inclined to the vertical at an angle θ and the extension of the string is a.**

   (a) **Find θ**

   (b) **Show that the modulus of elasticity of the string is $6mg\sqrt{2}$ newtons.**

(a)

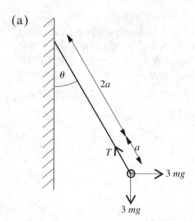

Considering the equilibrium of the particle:

Resolving $\leftarrow$ gives

$\qquad\qquad T \sin\theta - 3mg = 0 \qquad\qquad\qquad [1]$

Resolving $\uparrow$ gives

$\qquad\qquad T \cos\theta - 3mg = 0 \qquad\qquad\qquad [2]$

$[1] \div [2]$ gives

$\qquad\qquad \dfrac{T \sin\theta}{T \cos\theta} = \dfrac{3mg}{3mg} \qquad \Rightarrow \qquad \tan\theta = 1$

$\therefore \qquad\qquad \theta = 45°$

(b) The answer is required in surd form so we will express $\sin 45°$ in that way.

From [1], $\qquad\qquad T = \dfrac{3mg}{\sin 45°} = 3mg\sqrt{2} \qquad (\text{as } \sin 45° = \dfrac{1}{\sqrt{2}})$

Using Hooke's Law $\qquad T = \lambda\dfrac{a}{2a} \qquad \Rightarrow \qquad \lambda = 2T = 6mg\sqrt{2}$

The modulus of elasticity is $6mg\sqrt{2}$.

4. **Two identical springs, AC and BC, each of natural length a and modulus of elasticity $2mg$, are joined together at C. The ends A and B are attached to two points distant $4a$ apart vertically; A is above B. A particle of mass m is attached at C. Find the length of BC when the particle rests in equilibrium.**

   The extensions in the two springs are not equal as the weight of the particle is helping to stretch the upper spring but not the lower one.

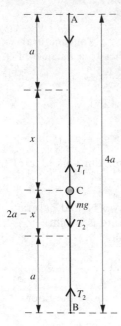

Taking x as the extension in AC we have $AC = a + x$.

Then $CB = 4a - (a + x) = 3a - x$

Therefore the extension in BC is $(3a - x) - a = 2a - x$

The particle is in equilibrium

∴ $T_1 - T_2 - mg = 0$ [1]

Using Hooke's Law

For AC $T_1 = \dfrac{\lambda x}{a} = \dfrac{2mgx}{a}$ [2]

For CB $T_2 = \dfrac{\lambda(2a - x)}{a} = \dfrac{2mg(2a - x)}{a}$ [3]

Using [2] and [3] in [1] we have

$$\dfrac{2mgx}{a} - \dfrac{2mg(2a - x)}{a} - mg = 0$$

∴ $\dfrac{4mgx}{a} = 5mg$ ⇒ $x = \tfrac{5}{4}a$

$BC = 3a - x = \tfrac{7}{4}a$

5. The diagram shows a uniform rod AB, of length 0.6 m and weight 40 N, with the
end A smoothly hinged to a wall. One end of a light elastic string is attached to B
and the other end to a fixed post. When the rod is in equilibrium it is inclined to
the wall at 60° and the string is at right-angles to the rod. Given that the natural
length of the string is 0.8 m and its modulus of elasticity is 32 N, find the length of
the string when the rod is in equilibrium.

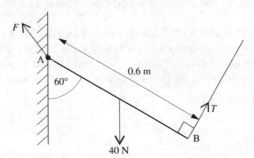

The hinge exerts an unknown force F on the rod at A. We can avoid having to find it if we take
moments about an axis through A.

A↻ gives $40 \times 0.3 \sin 60° - T \times 0.6 = 0$

⇒ $T = 20 \sin 60°$

Then using $T = \dfrac{\lambda x}{a}$ gives $20 \sin 60° = \dfrac{32x}{0.8}$ ⇒ $x = 0.433\ldots$

The length of the extended string is 1.23 m   (3 sf).

EXERCISE 1b

In questions 1 to 3 an elastic string, with modulus of elasticity λ N and natural length a m, has one end attached to a fixed point and the other to a particle of mass m kg which hangs in equilibrium.

1. $a = 0.8$, $m = 0.3$.   The string is stretched to length 1.3 m. Find λ.

2. $a = 0.7$, $m = 0.6$, $\lambda = 5$.   Find the extension of the string.

3. When $\lambda = 24$, the particle causes the string to stretch to three times its natural length. Find m.

In questions 4 to 6 a vertical spring, with modulus of elasticity λ N and natural length a m, has its lower end on the floor. On top of the spring is a light platform on which a particle of mass m kg rests in equilibrium.

4. $a = 0.3$, $m = 2$.   The particle is 0.2 m above the floor. Find λ.

5. $a = 0.5$, $m = 1.5$, $\lambda = 40$.   Find the height of the particle above the floor.

6. If $\lambda = 35$,   the particle compresses the spring to 60% of its natural length. Find m.

7. A spring is fixed at one end. When it hangs vertically, supporting a mass of 2 kg at the free end, its length is 3 m. The mass of 2 kg is then removed and replaced by a particle of unknown mass. The length of the spring is then 2.5 m. If the modulus of elasticity of the spring is 9.8 N, find the mass of the second load.

8. The end A of a light elastic string AB of natural length a and modulus of elasticity $2mg$ is fastened to one end of another light elastic string AC of natural length $2a$ and modulus of elasticity $3mg$. The ends B and C are stretched between two points $6a$ apart, so that BAC is a horizontal line.

   Find the length of AB.

9. A mass of 4 kg rests on a smooth plane inclined at 30° to the horizontal. It is held in equilibrium by a light elastic string attached to the mass and to a point on the plane. Find the extension in the string if it is known that a force of 49 N would double the natural length of 1.25 m.

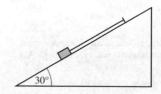

10. Two identical springs AC and BC of natural length a and modulus of elasticity $3mg$ are attached to a particle of mass m at point C. The ends A and B are attached to two points so that A is vertically above B and AB $= 3a$. The particle is in equilibrium between A and B. Find the length of AC.

11. The natural length of an elastic string AB is 2.5 m. The ends A and B are attached to two points on the same level and 2 m apart. A particle of mass 0.4 kg is attached to the mid point of the string and it hangs in equilibrium 1 m below the level of AB. Find the modulus of elasticity of the string.

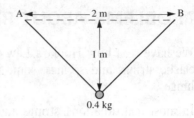

12. A light elastic string AB has natural length $2a$ and modulus of elasticity $mg\sqrt{3}$. The end A is attached to a fixed point and a particle of mass m is attached to B. The system is held in equilibrium, with B below the level of A and AB inclined at an angle θ to the vertical, by a force P at 90° to AB. In this position AB $= 3a$.

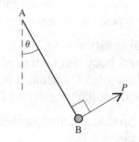

(a) Find θ.   (b) Find the force P.

13. A rod AB of mass m and length $4a$ is hinged at the end A to a fixed point. The end B is attached by an elastic string, of natural length $3a$, to point C, which is at a distance $4a$ vertically above A. The rod is in equilibrium at 60° to AC. Find

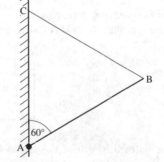

(a) the tension in the string

(b) the magnitude and direction of the reaction at the hinge

(c) the modulus of elasticity of the string.

14. ABCD is an elastic string of natural length 3 m and particles of equal mass are attached at B and C where AB $=$ BC $=$ CD $= 1$ m. The ends A and D are attached to two points on the same horizontal level and 3 m apart. The particles hang in equilibrium so that the string sections AB and CD are each at 60° to the horizontal.

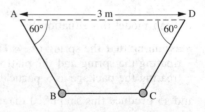

(a) Given that the extension in AB is x m, find an expression for the extension in BC.

(b) Find the value of x.

(c) The modulus of elasticity is 50 N. Find the mass of each particle.

PROBLEMS ABOUT REAL OBJECTS

We have seen how Hooke's Law can be used to solve problems involving light elastic strings and springs, some including *particles*, *uniform* rods and *smooth* hinges.

In most real situations, strings and springs do have some weight and objects to which they are attached may be quite large and of irregular shape. We can, however, *estimate* results by treating the real objects as though they were the ideal types, i.e. by assuming that:

real strings and springs are light and obey Hooke's Law,
a real body can be treated either as a particle with mass, or as light,
a long object, such as a plank, can be treated as a uniform rod,
pivots or hinges exert no friction.

This process is called *mathematical modelling* and is illustrated in the following example.

Example 1c

Raj is designing a machine to weigh heavy crates. A 'mock-up' of one of his designs uses a spring attached to a weighing platform as shown.

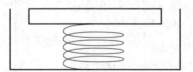

The natural length of the spring is 10 cm. When a crate, known to be of mass 100 kg is placed on the platform, it descends through 1 cm. How far should the platform descend when a mass of 150 kg is placed on it?

We can model this situation by

assuming that the spring obeys Hooke's Law,
treating the spring and the platform as light,
treating the package as a particle,

and so produce this simplified diagram.

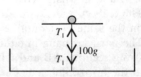

When the mass of 100 kg is in equilibrium, $T_1 = 100g$

The compression in the spring is 1 cm so Hookes Law gives

$$T_1 = \lambda \times \frac{1}{10} \quad \Rightarrow \quad \lambda = 1000g$$

If the thrust is T_2 when the mass of 150 kg is in equilibrium, $T_2 = 150g$

and Hooke's Law gives $T_2 = \dfrac{\lambda x}{10}$

$\therefore \quad 1500g = 1000gx \qquad \Rightarrow \qquad x = 1.5$

The estimated distance that a 150 kg mass would descend is 1.5 cm.

EXERCISE 1c

1. In a pinball machine the ball, of mass 30 g, is
   propelled from a cup, of mass 90 g, which is
   attached to a spring. The spring has modulus of
   elasticity 9 N and natural length 15 cm.

   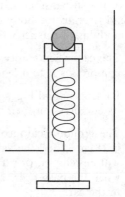

   (a) The mechanism allows the spring to be given
       a compression of 10 cm. Find the force
       necessary to do this.

   (b) Assuming that the masses of the spring and
       the holding mechanism are negligible, find
       the initial acceleration of the total mass of
       the cup and ball when the spring is released.

2. The spring for some kitchen scales is of length 10 cm. It is connected to a dial
   which is intended to measure loads up to 49 N. Design considerations suggest
   that the compression should not exceed 2.5 cm. By assuming that the scale pan is
   weightless find the minimum value you would recommend for the modulus of
   elasticity of the spring.

3. A door latch mechanism contains a spring to return the latch to the closed
   position. When the latch is in this position the spring is held compressed by
   means of a peg which prevents it from expanding. The spring has modulus of
   elasticity 12 N and natural length 8 cm.

   (a) In the closed position the spring exerts a force of 5 N on the peg. Find the
       compression of the spring.

   (b) In opening the door the spring is compressed by a further 1 cm. Find the
       force required to hold the latch in the open position.

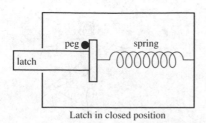

Latch in closed position

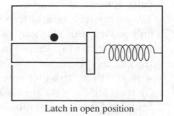

Latch in open position

4. 'Springmakers' Ltd. are testing springs for new application. They all have the same modulus of elasticity and are of various lengths.

   (a)   A spring of natural length 0.3 m is found to extend to 0.35 m when a force of 240 N is applied. Find the modulus of elasticity of the spring under test.

   (b)   In the application the spring will be subjected to tensions up to a value of 300 N. It is required that the extension should not exceed 0.05 m. Find the maximum length of spring which could be used.

5. Part of the suspension system on one rear wheel of a car consists of a spring with one end fixed to the body of the car and the other end fixed to the wheel axle. The modulus of elasticity of the spring is 8000 N.

   (a)   The car, when empty, puts a load of 2940 N on this spring, which compresses it to a length of 25 cm. Find the natural length of the spring.

   (b)   Passengers and luggage increase the load on this spring by 1370 N. What is its compression then?

   (c)   The ground clearance of the empty car is 18 cm. Find the value to which this will be reduced (for a stationary car) when the passengers and luggage are in the car.

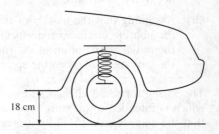

18 cm

6. Jan does a leg exercise. She raises her leg sideways to make an angle θ with the vertical, holds the position and then lowers the leg. Her legs are 70 cm long and the mass of one leg is 12 kg.

   (a)   Find the torque (i.e. the moment of the force) provided by her muscles when
(i) $\theta = 60°$ (ii) $\theta = 90°$.
State any assumptions you make in doing this.

   (b)   Jan's friend buys the 'Dinkylegs thigh trimmer', which consists of two ankle cuffs connected by an elastic string. The string has natural length 30 cm and modulus of elasticity 500 N. Find the total torque provided by her leg muscles when she does the exercise with the thigh trimmer when
(i) $\theta = 60°$ (ii) $\theta = 90°$.
State any further assumptions made.

7. An anti-vibration mount for a
machine contains a spring attached to
a fixed base plate and to a platform,
which can move vertically. A machine
of mass 240 kg has a rectangular base,
and a mount is placed under each
corner of this base. It is required
that, when the machine is not running,
the springs should be compressed to
half their natural length.

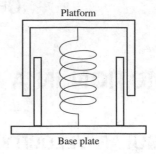

Platform

Base plate

(a) Find the thrust in each spring. State any assumptions you make in doing this.

(b) Find the modulus of elasticity of the springs.

*8. When a gardener uses his secateurs, his
hand extends over a section of each
handle between 5 cm and 13 cm from the
hinge. There is a spring between the
handles to return them to the open
position. This spring is attached to each
handle at a point 4 cm from the hinge. It
has a natural length 6 cm and modulus of
elasticity 60 N. The length of the spring
when the secateurs are fully open is 5 cm
and when they are closed it is reduced to
2 cm.

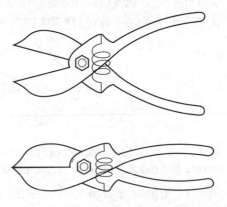

Given that there is nothing between the
blades, find

(a) the force that the gardener must exert when starting to close the handles

(b) the force needed to hold the secateurs in the fully closed position.

State all the assumptions you make in modelling this situation.

*9. In the diagram AB is a window hinged at A.
The stay of this window has broken and,
as a temporary replacement, a piece of
elastic rope of natural length 0.6 m, has been
attached to the window at B and to the
frame at D. This rope holds the window
at an angle of 50° to the vertical. The
weight of the window, 200 N, can be taken
as acting at its centre C.
AB = AD = 0.8 m. Find

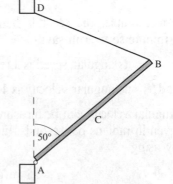

(a) the tension in the rope

(b) the modulus of elasticity of the rope.

CHAPTER 2

MOTION IN A HORIZONTAL CIRCLE

ANGULAR VELOCITY

Consider a particle that moves round the circumference of a circle with centre O. If the particle moves from a point P on the circumference to an adjacent point Q and the angle POQ is θ radians, then the rate at which θ is increasing is $d\theta/dt$. This is the *angular speed* of the particle and is often denoted by the symbol ω.

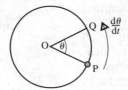

If we also state the direction of rotation (clockwise or anticlockwise), then we are giving the *angular velocity* of the particle.

We use a positive sign to denote the anticlockwise direction of rotation and a negative sign for clockwise rotation.

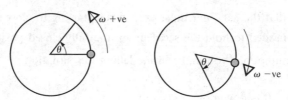

For example, the seconds hand of a clock rotates through 1 revolution in 1 minute so we can say:

its angular speed is $1 \, \text{rev} \, \text{min}^{-1}$

and   its angular velocity is $1 \, \text{rev} \, \text{min}^{-1}$ clockwise,   or $-1 \, \text{rev} \, \text{min}^{-1}$.

Angular velocity can be measured in revolutions per second, but is more usually given in radians per second. Either of these units can be converted to the other by using

$$1 \text{ revolution} = 2\pi \text{ radians.}$$

Constant Angular Velocity

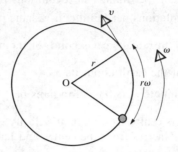

A particle that describes a circle at constant speed covers equal arcs in equal times so its angular velocity is also constant.

Suppose that the radius of the circle is r, the speed round the circumference is $v \, \mathrm{m\,s^{-1}}$ and the angular velocity is $\omega \, \mathrm{rad\,s^{-1}}$. In 1 second the particle travels round an arc of length v and this arc length is also given by $r\omega$, i.e.

$$v = r\omega$$

Examples 2a

1. Express   (a)   **3 revolutions per minute in radians per second**

              (b)   **0.005 radians per second in revolutions per hour.**

(a) $3 \, \text{rev/minute} = \dfrac{3}{60} \, \text{rev/second}$

$$= \frac{1}{20} \times 2\pi \, \text{rad s}^{-1} = \frac{\pi}{10} \, \text{rad s}^{-1}$$

(b) $0.005 \, \text{rad s}^{-1} = 0.005 \times 3600 \, \text{rad/hour} = 18 \, \text{rad/hour}$

$$= 18 \div 2\pi \, \text{rev/hour}$$

$$= \frac{9}{\pi} \text{rev/hour}$$

2. **A point on the circumference of a disc is rotating at a constant speed of $3 \, \mathrm{m\,s^{-1}}$. If the radius of the disc is $0.24 \, \mathrm{m}$ find, in $\mathrm{rad\,s^{-1}}$, the rate at which the disc is rotating.**

Using $v = r\omega$,   with $v = 3$ and $r = 0.24$ gives

$$3 = 0.24\omega \quad\quad \Rightarrow \quad\quad \omega = 12.5$$

$\therefore$ the disc rotates at $12.5 \, \mathrm{rad\,s^{-1}}$.

EXERCISE 2a

1. Express   (a)   0.2 radians per minute in revolutions per hour
             (b)   100 revolutions per minute in radians per second.

2. Find the angular velocity, in radians per second, of the minute hand of a clock.

3. Find the angular speed of the earth about its axis

   (a) in revolutions per minute   (b) in radians per second.

4. A disc is rotating about its centre with angular velocity ω rad s^{-1}. Point P is on the disc at a distance of d metres from the centre and has speed v metres per second.

   (a) $\omega = 6$, $d = 0.2$; find v.
   (b) $v = 5$, $d = 0.4$; find ω.
   (c) $v = 10$, $\omega = 2.5$; find d.

5. A fairground Big Wheel carriage is 8 m from the centre of the wheel, which is rotating at 10 revolutions per minute. Find the speed of the carriage.

6. Find the speed, in km h^{-1}, of a point on the equator of the earth, assuming the equator to be a circle of radius 6400 km.

7. A DIY power drill has a top rotational speed of 2500 revolutions per minute. A drill bit of diameter 3 mm has been inserted. Find the speed of the cutting edge of this drill.

8. A playground roundabout has a diameter of 3 m. A man puts his child onto it at a distance of 1 m from the centre and runs round pushing the edge of the roundabout with a speed of 2.4 m s^{-1}.

   (a) Find the angular speed in rad s^{-1}.
   (b) Find the speed with which the child is moving.

9. At a well a bucket is attached to a thin rope which is wound round a cylinder. The bucket is raised and lowered by turning the cylinder. The cylinder has a radius of 10 cm.

   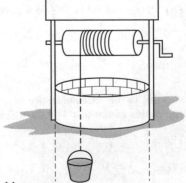

   (a) If the cylinder is turned at 1 revolution per second, at what speed, in m s^{-1}, will the bucket ascend?

   (b) The well owners decide to change the cylinder so that when it is turned at 1 rev/second, the bucket will ascend at 1 m s^{-1}. What should be the radius of the new cylinder?

   State any assumptions you make in solving this problem.

ACCELERATION

If the velocity of a moving object is changing, that object has an acceleration. As velocity is a vector, it can change in magnitude or in direction or both. It is easy to accept that changing *speed* involves acceleration but it not so easy to see that, for example, a car going round a corner at *constant* speed is accelerating because its direction is changing.

A change in speed is caused by a force that acts *in the direction* of motion of the object to which it is applied.

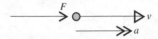

A force of this type cannot produce a change in the direction of the velocity so, if no other force is acting, the object continues to move in a straight line. The acceleration produced is a change in speed, i.e. a change in the magnitude of the velocity.

This type of acceleration was covered in detail in Module E.

A force that is perpendicular to the direction of motion of an object will push or pull the object off its previous line of motion but cannot alter the speed.

The acceleration in this case is a change in the *direction* of the velocity and the object therefore moves in a curve of some sort; the actual curve described depends upon the particular force acting.

A force that is neither parallel nor perpendicular to the direction of motion of an object has a component in each of these directions. Therefore it causes a change both in the speed and in the direction of motion of the object and the object moves with varying speed on a curved path.

MOTION IN A CIRCLE WITH CONSTANT SPEED

The direction of motion of a particle moving in a circle is constantly changing, so there must be a force acting perpendicular to the direction of motion of the particle at any instant.

If the particle is moving with constant speed there is no force acting in the direction of motion, i.e. no tangential force.

At any point on its path the particle is moving in the direction of the tangent at that point. A force that is perpendicular to this direction acts along the radius at that point. Further, because it is moving the particle from the tangent on to the circumference, the force must act *inwards* along the radius.

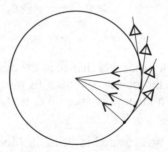

**The force that produces circular motion with constant speed
is at any instant acting radially inwards on the particle,
producing a radial acceleration.**

The Magnitude of the Radial Acceleration

Consider a particle moving at constant speed v, round a circle of centre O and radius r. Suppose that, in a time δt, the particle moves from a point P to a nearby point Q, through a small angle $\delta\theta$ measured in radians. For reasons of clarity, $\delta\theta$ is not drawn as a very small angle in the diagram below.

The length of the arc PQ is $r\delta\theta$ and, as this arc is covered in time δt at speed v, its length is also $v\delta t$.

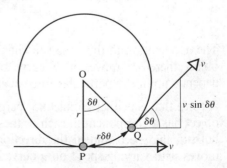

Therefore $r\delta\theta = v\delta t$

$\Rightarrow$ $\dfrac{v}{r} = \dfrac{\delta\theta}{\delta t}$

When the particle is at P it has no velocity component along PO, but when it reaches Q the velocity component parallel to PO is $v \sin \delta\theta$.

Therefore the average acceleration from P to Q, in the direction of PO, is given by

$$\frac{v \sin \delta\theta}{\delta t}$$

The closer Q is to P, the nearer this expression is to the actual radial acceleration at P and, at the same time, both $\delta\theta$ and δt approach zero. When this happens the value of $\dfrac{v \sin \delta\theta}{\delta t}$ becomes indeterminate.

However, for any angle α measured in radians, it can be shown that

$$\text{as} \quad \alpha \to 0, \quad \sin\alpha \to \alpha$$

A formal proof of this property is not given at this stage but you will probably find it quite convincing to compare the values of small angles and their sine ratios using a calculator.

Therefore as $\quad \delta\theta \to 0 \qquad \sin \delta\theta \to \delta\theta$

$$\therefore \qquad\qquad \frac{v \sin \delta\theta}{\delta t} \quad \to \quad v\frac{\delta\theta}{\delta t} \quad \to \quad v\frac{d\theta}{dt}$$

But $\dfrac{d\theta}{dt}$ is the angular velocity, ω.

Hence the radial acceleration of the particle is $v\omega$.

Then using $v = r\omega$, this acceleration can be expressed as either $r\omega^2$ or $\dfrac{v^2}{r}$.

To sum up:

**The acceleration of a particle travelling in a circle of radius r,
at constant speed v (or constant angular speed ω),
is directed *towards* the centre of the circle
and is of magnitude $r\omega^2$ or $\dfrac{v^2}{r}$.**

It follows that

**a particle can describe a circle with constant speed
only when under the action of a force of constant magnitude
directed *towards the centre* of the circle.**

Examples 2b

1. A particle P is travelling round a circle of radius 0.8 m at a constant speed of 2 m s^{-1}. Find the acceleration of the particle, giving its magnitude and direction.

The acceleration is given by $\dfrac{v^2}{r}$, where $v = 2$ and $r = 0.8$

The magnitude of the acceleration is $\dfrac{2^2}{0.8}$ m s^{-2}, i.e. 5 m s^{-2}, and it is directed towards the centre of the circle.

2. A firework consists of a strip of wood, pivoted at its centre, with a rocket attached to each end. The length of the strip is 0.9 m. After the rockets are lit the firework rotates at 12 rev s^{-1}. Use a suitable model to estimate the radial acceleration of each rocket.

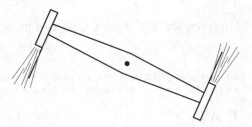

Model each rocket as a particle at a distance 0.45 m from the pivot.

The angular speed is 12 rev s^{-1}, i.e. $12 \times 2\pi$ rad s^{-1}

The acceleration of one particle is given by $r\omega^2$

When $r = 0.45$ and $\omega = 24\pi$, $r\omega^2 = (0.45)(24\pi)^2 = 2560$ (3 sf)

An estimate of the radial acceleration of each rocket is 2560 m s^{-2} towards the pivot.

EXERCISE 2b

In questions 1 to 5 a particle is travelling round a circle, centre O, radius r metres, at a constant speed of v m s^{-1} and angular velocity ω rad s^{-1}. Its acceleration is of magnitude a m s^{-2}.

1. $v = 16$, $r = 5$; find a.

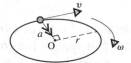

2. $\omega = 12$, $r = 3$; find a.

3. In what direction is the acceleration

   (a) when the particle has reached a point P on the circle

   (b) when the particle has reached a point Q on the circle.

4. $a = 75$, $r = 12$; find v.

5. $a = 500$, $\omega = 6$; find r.

6. Taking the earth to be a sphere of radius 6400 km, calculate the acceleration, in ms^{-2}, due to the earth's rotation of

   (a) a person A who is standing on the equator

   (b) a person B who is at a point which is 3200 km from the earth's axis of rotation.

   Show on a diagram the direction of the acceleration in each case.

7. An aircraft waiting to land at a busy airport is circling at a constant height, at 500 km h^{-1}. The passengers experience an acceleration of $0.4g$ m s^{-2}. Find the radius of the circle.

8. A machine is designed to test astronauts in conditions of great acceleration. The astronaut is strapped into a chair which is then moved round in a horizontal circle of radius 5 m. If he can withstand accelerations up to $9g$ m s^{-2}, what is the maximum permissible angular velocity?

9. Passengers on a fairground ride are whirled round in a horizontal circle of radius 6 m, experiencing a radial acceleration of 15 m s^{-2}. At what speed are they moving?

10. A quarter-scale railway line is to be laid for a model train. It has been decided that the children who ride on the train should not be subjected to a radial acceleration greater than 8 m s^{-2}. By treating the bends in the line as arcs of circles, find the speed limit you would recommend if the radius of the sharpest bend is 10 m.

MOTION IN A HORIZONTAL CIRCLE

We know that when a circle is described at constant speed, there is an acceleration of constant magnitude v^2/r or $r\omega^2$ towards the centre of the circle and no acceleration tangentially. There must therefore be

(a) no tangential force acting

(b) a force of constant magnitude acting towards the centre.

These conditions can be achieved when the circle is in a horizontal plane because the weight of the particle, being a vertical force, has no component in the direction of motion of the particle. So it is possible for the circular motion to be performed at a constant speed. (In a vertical plane, on the other hand, as a particle moves in a circular path the tangential component of the weight varies, causing the speed of the particle to vary.)

There are many ways in which the necessary force towards the centre can be provided, e.g. by a rotating string with one end fixed at the centre or by friction with the road surface as a car turns round a bend. Some of the possibilities are illustrated in the following examples.

Examples 2c

1. One end of a light inelastic string of length a is fixed to a point O on a horizontal plane. A particle P of mass m, attached to the other end of the string, is given a blow which sets it moving in a circle on the plane, with constant angular velocity ω.

   (a) Find the tension in the string and the force exerted on P by the table.

   (b) Explain an assumption that has been made in the question about a certain possible force.

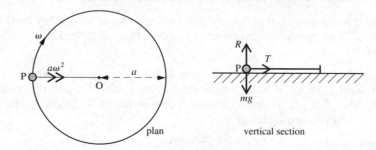

   plan   vertical section

Vertically there is no acceleration; horizontally there is an acceleration of $a\omega^2$ towards O.

   (a) Resolving ↑ $R - mg = 0$ [1]

   Using Newton's Law → $T = ma\omega^2$ [2]

   The tension is $ma\omega^2$ and the reaction exerted by the table is mg.

   (b) As P travels with constant speed it is assumed that there is no tangential force, i.e. that there is no friction between the particle and the plane.

2. A small block A, of mass m kg, lies on a horizontal disc which is rotating about its centre B at $3 \, \text{rad s}^{-1}$ and A is 0.8 m from B. If the block does not move relative to the disc, find the least possible value of μ, the coefficient of friction, between the block and the disc.

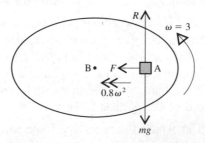

The frictional force F acing on A is towards the centre of the circle because A has to be given a central acceleration in order to travel in a circle; there is no friction tangentially because A has no tendency to move in that direction. Also, as the block is small we treat it as a particle.

Resolving ↑ $\qquad\qquad R - mg = 0 \qquad\qquad \Rightarrow \qquad R = mg$

Using Newton's Law ← $\qquad F = mr\omega^2 \qquad \Rightarrow \qquad F = m(0.8)(3)^2 = 7.2m$

Now $\quad F \leqslant \mu R \qquad \Rightarrow \qquad 7.2m\omega^2 \leqslant \mu mg$

i.e. $\qquad\qquad\qquad\qquad\qquad\qquad \mu \geqslant \dfrac{7.2}{g}$

∴ the least value of μ is $\dfrac{7.2}{9.8}$ i.e. 0.735 (3 sf).

3. Katie is hoping to prepare her pony, Ben, for showing in a ring. As part of the training programme she is holding one end of an extensible rope and the other end is fastened to Ben's bridle. The rope has an unstretched length of 8.6 m and a modulus of elasticity 740 N.

(a) Ben is very obediently trotting at a steady speed of $2.5\,\mathrm{m\,s^{-1}}$ in a circle of radius 11 m.
    Find　(i) the tension in the rope　(ii) Ben's mass.

(b) The rope ceases to obey Hooke's Law if the extension exceeds 3 m.
    Find the greatest speed that Ben should be asked to achieve.

We will model Ben as a particle, the rope as a light elastic string, and Katie as a point at the centre of the circle. As these assumptions are fairly rough, answers corrected to 3 sf are not appropriate so we will use only 2 sf.

(a)

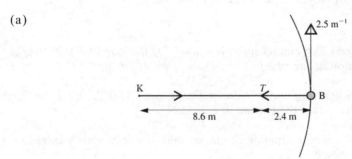

(i) The rope is extended by 2.4 m therefore Hooke's Law, $\quad T = \dfrac{\lambda x}{a} \quad$ gives

$$T = \frac{740 \times 2.4}{8.6} = 206.5\ldots$$

The tension in the rope is 210 N (2 sf).

(ii) Ben's acceleration towards K is given by $\dfrac{v^2}{r}$, so it is $\dfrac{6.25}{11}\,\mathrm{m\,s^{-2}}$.

Now using Newton's Law, $\quad T = ma$, gives

$$206.5\ldots = m \times \frac{6.25}{11} \qquad \Rightarrow \qquad m = 360 \quad (2\ \mathrm{sf})$$

Ben's mass is 360 kg (2 sf).

(b)

$$K \xrightarrow{\quad T \quad} \xleftarrow{\quad T \quad} B \quad \uparrow v$$

8.6 m   3 m

When the extension is 3 m, the tension becomes $\dfrac{740 \times 3}{8.6}$ N

Then using $T = \dfrac{mv^2}{r}$ gives $v^2 = \dfrac{Tr}{m} = \dfrac{740 \times 3 \times (8.6 + 3)}{8.6 \times 363}$

$\Rightarrow$ $v = 2.87\ldots$

It is dangerous to give the speed *corrected* to 2 sf, as that increases the result slightly. So instead we will *truncate* the calculated figure to 2 sf.

Ben should not exceed a speed of $2.8\,\mathrm{m\,s^{-1}}$.

EXERCISE 2c

In questions 1 and 2, one end A of an inelastic string AB is fixed to a point on a smooth table. A particle P is attached to the other end B, and moves on the table in a horizontal circle with centre A.

1. The particle is of mass 1.5 kg and its speed is $4\,\mathrm{m\,s^{-1}}$. If the length of the string is 2.4 m, find the tension in the string.

2. The mass of the particle is 8 kg and it is moving with speed $5\,\mathrm{m\,s^{-1}}$. Find the length of the string given that the tension in it is 12 N.

   Questions 3 and 4 are about a situation similar to that described above, except that the string AB is elastic.

3. The elastic string AB has a natural length of 2.5 m and its modulus of elasticity is 40 N. The mass of P is 5 kg. If the string is extended by 0.5 m, find the speed of P.

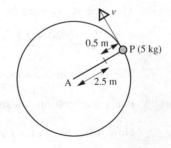

4. When P has a mass of 1.5 kg, and is moving with a speed of $6\,\mathrm{m\,s^{-1}}$, the string is extended by 0.4 m. Given that the natural length of the string is 2 m, find its modulus of elasticity.

5. A circular tray of radius 0.2 m has a smooth vertical rim round the edge. The tray is fixed on a horizontal table and a small ball of mass 0.1 kg is set moving round the inside of the rim of the tray with speed $4 \, \text{ms}^{-1}$. Calculate the horizontal force exerted on the ball by the rim of the tray.

6. A particle of mass 0.4 kg is attached to one end of a light inextensible string of length 0.6 m. The other end is fixed to a point A on a smooth horizontal table. The particle is set moving in a circular path.

    (a) If the speed of the particle is $8 \, \text{m s}^{-1}$ calculate the tension in the string and the reaction with the table.

    (b) If the string snaps when the tension in it exceeds 50 N, find the greatest angular velocity at which the particle can travel.

    The remaining questions in this exercise involve choosing a model. This should be clearly described and all assumptions mentioned.

7. An aircraft is flying at $700 \, \text{km h}^{-1}$ in a horizontal circle of radius 2 km.

    (a) Find the horizontal and vertical components of the thrust exerted by the seat, on a passenger of mass 60 kg.

    (b) Find the magnitude and direction of the resultant thrust.

8. A satellite, of mass m kilograms, is orbiting the earth on a circular path at a height of 100 km above the surface. At this height the acceleration due to gravity is $9.5 \, \text{m s}^{-2}$. Take the radius of the earth as 6400 km. Find

    (a) the force exerted on it towards the centre of the earth

    (b) its speed in m s^{-1}

    (c) its angular speed

    (d) the time it takes to perform one orbit.

THE CONICAL PENDULUM

Consider a light inelastic string fixed at one end A and carrying a particle hanging freely at the other end. If the particle, which is not resting on a surface of any sort, is set moving in a horizontal circle, the plane of that circle will be below the level of A. As the particle and the string rotate, they trace out the surface of a cone and the system is known as a *conical pendulum*.

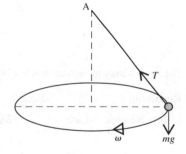

In this situation the tension in the string does two jobs – the horizontal component provides the central force needed to keep the particle rotating and the vertical component balances the weight of the particle. It follows that the string can never be horizontal, as the tension in it *must* have a vertical component.

There is a similar situation when a particle moves on the inner surface of a smooth sphere, in a horizontal circle below the level of the centre of the sphere.

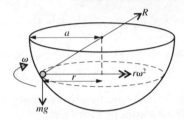

The vertical component of the normal reaction R (which acts through the centre of the sphere) balances the weight of the particle, while the horizontal component of R acts on the particle towards the centre of the circle being described.

Examples 2d

1. A particle P, of mass m, is attached to one end of a light inextensible string of length a and describes a horizontal circle, centre O, with constant angular speed ω. The other end of the string is fixed to a point Q and, as P rotates, the string makes an angle θ with the vertical. Show that

    (a) the tension in the string always exceeds the weight of the particle,

    (b) the depth of O below Q is independent of the length of the string.

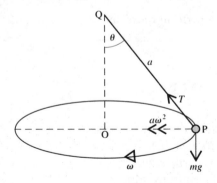

The string cannot be vertical therefore $\theta > 0$.

Resolving ↑ $\qquad\qquad\qquad T\cos\theta - mg = 0$ $\qquad\qquad\qquad\qquad$ [1]

Using Newton's Law ← $\qquad\quad T\sin\theta = m(a\sin\theta)\omega^2$

$\Rightarrow \qquad\qquad\qquad\qquad\qquad T = ma\omega^2$ $\qquad\qquad\qquad\qquad\qquad$ [2]

(a) From [1], $T = \dfrac{mg}{\cos \theta}$ and $\cos \theta < 1$

$\therefore$ $T > mg$

(b) In $\triangle$POQ, $QO = a \cos \theta$ [3]

From [1] and [3] $QO = \dfrac{amg}{T}$

Then from [2] $QO = \dfrac{amg}{ma\omega^2} = \dfrac{g}{\omega^2}$

i.e. the depth of O below Q is independent of the length of the string.

2. **An elastic string, of natural length l and modulus of elasticity $2mg$, has one end fixed to a point A and has a particle P, of mass m, attached to the other end B. P is travelling in a horizontal circle with angular speed ω. The string reaches its elastic limit when the tension is $3mg$. Find the angular speed when this state is reached.**

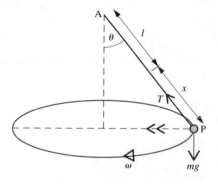

$$T = 3mg \qquad [1]$$

Resolving $\uparrow$ $T \cos \theta - mg = 0$ [2]

Newton's Law $\leftarrow$ $T \sin \theta = m\omega^2 (l + x) \sin \theta$ [3]

Hooke's Law, $T = \dfrac{\lambda x}{l} \Rightarrow T = \dfrac{2mgx}{l}$ [4]

From [1] and [4] $3mg = \dfrac{2mgx}{l} \Rightarrow x = \tfrac{3}{2}l$

From [1] and [3] $3mg = m\omega^2(1 + x) \Rightarrow 3g = \omega^2\left(\tfrac{5}{2}l\right)$

Therefore $\omega = \sqrt{\dfrac{6g}{5l}}$

Note that one of the equations formed was not used. This happens quite often but it is not easy to spot at the outset. So, unless you are *sure* that an equation will not be needed, the best policy is to apply all the relevant principles and then see which equations are useful.

3. A particle P is moving on the inner surface of a smooth hemispherical bowl with centre O and radius $2a$. The particle is describing a horizontal circle, centre C, with angular speed $\sqrt{\dfrac{g}{a}}$.

Find

(a)   the magnitude of the force exerted on P by the surface of the bowl

(b)   the depth of C below O.

(a)   For the forces acting on P,

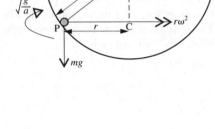

      Resolving ↑ gives

$$R \cos \theta - mg = 0 \qquad [1]$$

      Newton's Law → gives

$$R \sin \theta = mr\omega^2$$
$$= m(2a \sin \theta)\left(\frac{g}{a}\right)$$
$$\Rightarrow \qquad R = 2mg \qquad [2]$$

The force exerted on P is $2mg$.

(b)   In △OPC, $OC = 2a \cos \theta$

From [1] and [2], $\cos = \tfrac{1}{2}$

∴ $OC = a$

The depth of C below O is a.

4. A smooth ring R, of mass m, is threaded on to a light inextensible string of length 1.4 m. The ends of the string are fixed to two points A and B, distant 1 m apart in a vertical line. When the ring is set rotating in a horizontal circle with angular speed ω radians per second, the distance of the ring from the upper fixed point, A, is 0.8 m.

(a)   Show that ARB is a right angle and hence write down the values of $\sin \theta$ and $\cos \theta$ where θ is ABR.

(b)   Find the radius, r m, of the horizontal circle.

(c)   Find the value of ω, corrected to 2 sf.

(a)   BR $= (1.4 - 0.8)$ m $= 0.6$ m

∴   triangle ARB is a '3, 4, 5' triangle
with a right angle at R.

From triangle ABR, $\sin \theta = 0.8$
and $\cos \theta = 0.6$.

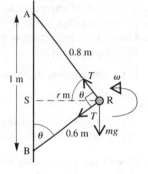

(b)   From triangle ARS, $r = 0.8 \cos \theta$

The radius of the circle is 0.48 m.

(c)   Using Newton's Law towards the centre of
the circle gives

$$T \cos \theta + T \sin \theta = m \times 0.48\omega^2$$

$\Rightarrow$ $1.4T = 0.48m\omega^2$ [1]

Vertically the ring is in equilibrium.

Resolving ↑ gives $T \sin \theta - T \cos \theta - mg = 0$

$\Rightarrow$ $0.2T = mg$ [2]

[1] ÷ [2] gives $\dfrac{1.4T}{0.2T} = \dfrac{0.48m\omega^2}{mg}$

$\Rightarrow$ $\omega^2 = \dfrac{7g}{0.48} = 1.429\ldots$

∴ $\omega = 12$ (2 sf)

Note.   In this problem the ring is smooth and is *threaded* on to the string;
therefore the tension is the same on both sides of the ring.

If, instead, the ring is *fastened* to a point on the string, the tensions in the two
portions of the string are, in general, different.

EXERCISE 2d

Questions 1 to 3 are about a conical pendulum which consists of an inextensible
string AB with a particle P attached at B. Point A is fixed and B moves in a
circle in a horizontal plane.

1.   The length of the string AB is 1.5 m and
the mass of the particle P is 3 kg. Given
that P is rotating in a circle with an
angular speed of 8 rad s^{-1} find

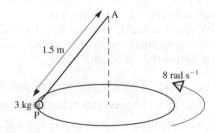

(a)   the tension in the string

(b)   the angle between the string and
the vertical.

2. The mass of the particle P is 2 kg and P
   is rotating in a circle of radius 0.3 m.
   Given that the string is inclined at 25° to
   the vertical find

   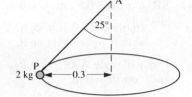

   (a) the tension in the string

   (b) the angular speed of P.

3. The particle P is rotating in a circle with an angular speed of 5 rad s^{-1}. Find the
   depth below A of the plane of this circle.

   Questions 4 and 5 are about a situation similar to that described for questions 1
   to 3, except that the string AB is elastic.

4. The mass of the particle P is 2 kg. The
   elastic string AB has a natural length of 0.4 m
   and its modulus of elasticity is 12 N.
   Given that when P rotates in a circle the
   extension of the string is 0.1 m find

   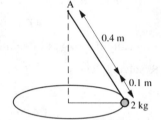

   (a) the tension in the string

   (b) the angular speed of P.

5. The mass of the particle P is 0.5 kg and P is rotating in a circle with an angular
   speed of 2 rad s^{-1}. The modulus of elasticity of the string AB is 3 N.
   Given that the string is inclined at 60° to the vertical find

   (a) the tension   (b) the extended length   (c) the natural length.

6. A particle of mass m, attached to the end A of a light inextensible string
   describes a horizontal circle on a smooth horizontal plane with angular speed ω.
   The string is of length $2l$ and the other end B is fixed,

   (a) to a point on the plane

   (b) to a point which is at a height l above the plane.

   Find, in each case, the tension in the string and the reaction between the particle
   and the plane, giving your answers in terms of m, l, g and ω.

7. One end of a light inextensible string of
   length $3a$ is attached to a fixed point A,
   and the other end to a point B which is
   at a distance $2a$ vertically below A.
   A small bead, P, of mass m, is fastened
   to the midpoint of the string and moves in
   a horizontal circle with speed $\sqrt{5ga}$.

   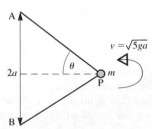

   (a) (i) Find the sine of the angle θ between the string and the horizontal.

       (ii) Express the radius of the circle in terms of θ.

   (b) Find the tensions in the two halves of the string.

8. Suppose that, in question 7, the bead is threaded on to the string (not fastened) and moves in a horizontal circle with centre R. The string is taut and $BR = \frac{5}{6}a$. Find the speed of P.

9. A boy has a ball of mass 0.15 kg on an elastic string of natural length 30 cm. He whirls the ball in a horizontal circle of radius 25 cm with the string at 50° to the vertical. Find

   (a) the tension in the string

   (b) the speed of the ball

   (c) the extended length of the string

   (d) the modulus of elasticity of the string.

10. A powered model aircraft of mass 0.25 kg is attached by a wire of length 10 m to a point A on the ground. The plane flies in a horizontal circle, of radius 5 m, centred above the point A. It is rotating at 12 revolutions per minute. The motion through the air produces a vertical lift force, P newtons, on the plane. Find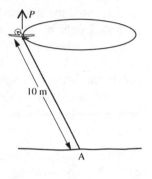

    (a) the angle between the wire and the ground

    (b) the tension in the wire

    (c) the lift force P.

11. Two boys decide to add interest to a train journey by doing an experiment with a pendulum, which they hang from the luggage rack. When the train has constant velocity the pendulum hangs vertically, as shown in the diagram. When the train goes round a bend at $v\,\mathrm{m\,s}^{-1}$, the pendulum deviates through an angle θ from this position.

    Consider the bend to be an arc of a circle. Show on a diagram the position of the pendulum when the train is going round the bend shown.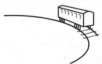

    The boys can calculate θ by measuring the distance of the pendulum bob, B, from the side of the train. They intend to estimate the radius, r metres, of each bend from a map and use r and θ to calculate v.

    (a) Obtain a formula for v in terms of r and θ.

    (b) On one bend they observe that $\theta = 5°$ and estimate r as 800 m. Find, in $\mathrm{km\,h}^{-1}$, the value they obtain for the speed of the train.

    (c) If the train accelerates on a straight section of track, describe how the pendulum behaves.

PRACTICAL PROBLEMS

It is not every day that most people see a particle at the end of a string rotating in a horizontal circle; a much more common sight is a vehicle turning round a corner at a roughly constant speed.

By modelling the vehicle as a particle, the speed as constant, the path taken as part of a circle and the road surface as level, we can find an approximation for the frictional force needed between the tyres and the road surface.

We have seen that it is possible to find estimates for the values of quantities in many practical problems by forming a mathematical model. Whether or not these estimates are reasonable depends upon the validity of the assumptions used to produce the model. In many cases the validity of the assumptions can be tested only by experiment. In other cases, common sense is often a good judge of whether a particular assumption is reasonable. For example, modelling a lorry as a particle is sensible if it is going round a wide bend but not if the bend is a tight one (watch an articulated lorry turn a corner and you will see that different wheels follow different paths round the bend). Whatever the case it is important to describe the model carefully, giving all the assumptions that are made.

When a practical problem has been modelled, it becomes a simplified mathematical exercise. The working diagram we use in the solution no longer needs to be realistic; instead of large objects such as cars, trees, large springs, planks etc, it is made up of points (i.e. particles) and lines. Then when forces, velocities, dimensions and so on are marked on this diagram, they can be seen clearly.

In this type of problem, where the assumptions made are inevitably some way from the actual situation, it is inappropriate to give answers to more than two significant figures.

Example 2e

A car of mass $600\,\text{kg}$ turns left at constant speed $4\,\text{m}\,\text{s}^{-1}$, moving on a circular path of radius $4.7\,\text{m}$.

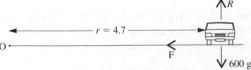

(a) Find the least value of the coefficient of friction between the road and the tyres if the bend is to be taken without skidding.

(b) State, with comments on their validity, any assumptions that you have made in your solution.

(a)   As the car moves with constant speed while turning the corner, there is no overall force acting in the direction of motion. There is a frictional force on the tyres towards the centre of the circular path.

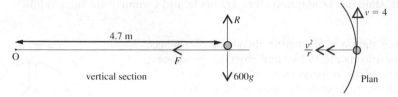

Resolving ↑ $R - 600g = 0$ $\Rightarrow$ $R = 600g$ [1]

Newton's Law ←, $F = \dfrac{mv^2}{r}$ $\Rightarrow$ $F = \dfrac{(600)\,(4^2)}{4.7}$ [2]

Using $F \leqslant \mu R$ gives

$$\frac{600 \times 16}{4.7} \leqslant 600 \times 9.8 \times \mu \quad \Rightarrow \quad \mu \geqslant 0.35 \quad (2\ \text{sf})$$

The minimum value of μ is 0.35   (2 sf).

(b)   Assumptions made are:

The car is small enough to be treated as a particle – this is a substantial assumption as the radius of the corner might not be much greater than the length of the car.

The road surface is level – not unreasonable but unlikely to be true as roads are usually cambered.

No other forces, such as air resistance, affect the motion – although this can never be *true*, it is not unreasonable because its effect on low-speed motion is small.

Note that the frictional forces between the tyres of a road vehicle and the road surface are a complex issue which has been greatly simplified in the example above.

The driving force of the vehicle causes the wheels to rotate so as to drive the car forward but this is possible only if there is a frictional force that prevents the tyres from spinning on the spot, i.e. there *is* some friction along the line of motion.

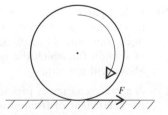

What we can be sure of however is that, at constant speed, the *resultant* force in the direction of motion is zero.

EXERCISE 2e

In this exercise use a mathematical model to solve the problems. Describe the model, state the assumptions that are made and comment on their validity.

1. A space station is planned in the shape of a wheel. The astronauts are to live and work in the space corresponding to the position of a tyre. The station has an overall diameter of 50 m. To provide a simulation of gravity the station will be made to rotate about its centre C. Find the angular velocity necessary to produce an acceleration of $9.8 \, \mathrm{m \, s^{-2}}$ at a point 25 m from the centre, e.g. point P on the diagram.

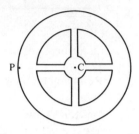

2. An object of mass 0.3 kg is placed on a horizontal rotating turntable at a distance of 0.2 m from the axis. The coefficient of friction between the object and the turntable is 0.4. Find the greatest possible value for the angular velocity if the object is not to slip outward from its position.

3. A car of mass 800 kg is travelling round a bend of radius 150 m.

   (a) Find the frictional force on the tyres when the speed is $30 \, \mathrm{m \, s^{-1}}$.

   (b) The coefficient of friction is 0.7. Find the maximum speed at which the car can go without skidding outwards.

4. A van is carrying a parcel of mass 5 kg. When the van goes round a corner the parcel, provided that it does not slip, follows a path which is to be treated as an arc of a circle.

   (a) If the radius of the arc on which the parcel moves is 3.8 m and the van is cornering at $30 \, \mathrm{km \, h^{-1}}$, find the frictional force on the parcel, assuming that it does not slip.

   (b) In fact parcels have been sliding about and it is decided to provide a rougher surface inside the van. Find the least coefficient of friction for which the parcel described above will not slip.

   (c) A surface coating is provided for which the coefficient of friction is 2. If the driver takes a bend at a speed of $50 \, \mathrm{km \, h^{-1}}$, with the parcel moving on an arc of radius 10.5 m, will the parcel now slip?

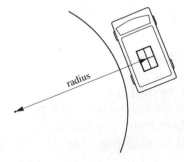

radius

5. A ride at a funfair consists of cars which are made to move in a horizontal circle, of radius 4 m, at a rate of 0.3 revolutions per second. A girl of mass 40 kg is riding in a car. Find the horizontal and vertical components of the force exerted by the car on the girl.

6. On a fairground ride, people stand against the sides of a cylinder, of radius r metres, which then starts to rotate about its axis which is vertical. When a suitable angular velocity ω rad s^{-1} is reached, the floor descends and the people are held in place by the friction between them and the wall. The coefficient of friction is μ. A man of mass m kilograms takes the ride.

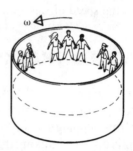

(a) Find the normal reaction force from the wall on the man.

(b) Find the frictional force necessary to prevent him from sliding down the wall.

(c) Find the least angular speed at which the floor can be lowered.

(d) Evaluate this angular speed for the case $r = 2.5$, $\mu = 0.4$, and find the speed with which the man is then moving.

7. Aaron, Beth, Carol and Dipak are skating. They hold hands, in this order, with their arms outstretched. They are of similar size. Their average span from left hand to right hand is 150 cm and each of their masses is approximately 50 kg. Aaron stays on a spot and the others skate round him in circles, with the same angular velocity and staying in a straight line along a radius.

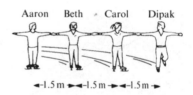

(a) If the angular velocity is ω rad s^{-1}, find the speeds of Beth, Carol and Dipak in terms of ω.

(b) By modelling the skaters as particles and assuming that they are not using their skates to provide any force towards the centre, find in terms of ω,

(i) the force T_1 which Carol exerts on Dipak

(ii) the force T_2 which Beth exerts on Carol

(iii) the force T_3 which Aaron exerts on Beth.

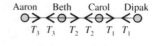

8. A girl is swinging on a rope, of length 5 m, attached to a swivel on top of a pole. It takes her 4 s to complete a circle around the pole. Her mass is 40 kg. Find

(a) the tension in the rope

(b) the angle between the rope and the pole

(c) the radius of the circle.

9. A small bead B of mass 0.2 kg is rotating in a horizontal circle on the inner surface of a smooth hemisphere of radius 0.3 m and centre O. If the centre of the horizontal circle is 0.1 m below O, find

(a)   the magnitude of the force exerted by the bowl on the bead,

(b)   the speed of the bead.

BANKED TRACKS

When racing cars, cycles, motorbikes, etc. are rounding the bends of a track at high speeds, the available frictional force alone is unlikely to be sufficient to prevent the car from slipping sideways on the track so a further central force has to be provided. This is done by banking the track, i.e. by raising the outside of the curve above the level of the inside. This has the effect of 'tipping' the normal contact force that the ground exerts on the vehicle, away from the vertical so that it has a horizontal component. This component then forms part of the resultant force towards the centre. At the same time, however, only the horizontal component of friction now acts along the radius.

Consider a car of mass m kg, racing on a track that has curved bends with a radius of r metres. The curved sections are banked at an angle θ (i.e. at θ to the horizontal).

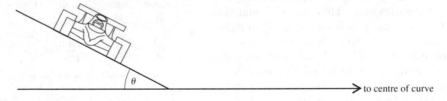

If the car travels round the bends at a speed of v metres per second, such that a frictional force acts down the slope, the following relationships can be formed.

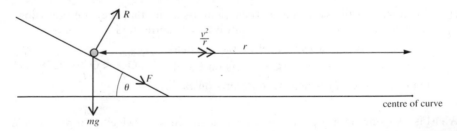

Resolving ↑ $R \cos \theta - F \sin \theta - mg = 0$

Newton's Law → $R \sin \theta + F \cos \theta = mv^2/r$

Also $F \leqslant \mu r$

These equations provide the means of solving various problems, e.g. the maximum possible speed for a given angle of banking can be found.

In particular the angle at which the track should be banked for a specified design speed (i.e. the speed at which there is no tendency to slip and therefore no frictional force is needed), can be estimated.

Note that any such values can only be estimates, as a number of assumptions have been made, e.g. the car is modelled as a particle;
the difference between the outer and inner radius is ignored;
constant speed is assumed round the curve.

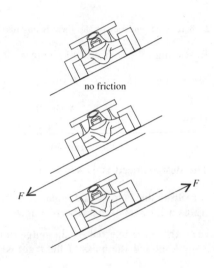

When a car moves at the design speed it has no tendency to slip sideways on the track so there is no friction up or down the banked track.

If the car's speed is higher than the design speed it tends to slip *up* the track and friction therefore acts *down* the track, i.e. it has a component *towards* the centre of the curve.

If the car's speed is *lower* than the design speed it tends to slip *down*, and friction therefore acts *up* the track, i.e. it has a component *away from* the centre of the curve.

Bends on a railway track are dealt with in a similar way, the outer rail being raised above the level of the inner rail. The difference is that no friction is involved in this case, as movement up or down the banked track is prevented by the lateral force exerted by the rails on the flanges of the wheels.

tendency to   tendency to
move inwards   move out

In the case of trains moving round banked curved tracks it is much more reasonable to assume constant speed and to ignore the difference in radii as either radius is very large.

Examples 2f

1. A curved section of a race track, where the radius is 120 m, is banked at 40°. By modelling a car that drives round the track as a particle, show that the design speed for this section, $V \mathrm{m\,s^{-1}}$, is independent of the mass of the car and find its value.

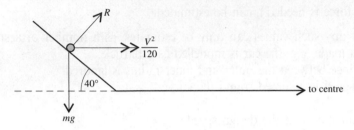

If there is no tendency to slip there is no lateral frictional force.

Resolving $\uparrow$ $\qquad R \cos 40° - mg = 0$ [1]

Newton's Law $\rightarrow$ $\qquad R \sin 40° = m \dfrac{V^2}{120}$ [2]

Hence $\qquad \dfrac{R \sin 40°}{R \cos 40°} = \dfrac{mV^2}{120} \div mg = \dfrac{V^2}{120g}$

$\therefore \qquad V^2 = 120g \times \tan 40°$ which is independent of m.

The design speed is $31.4 \mathrm{m\,s^{-1}}$ (3 sf) i.e. $113 \mathrm{km\,h^{-1}}$ (3 sf)

2. A railway line is to be laid round a circular arc of radius 500 metres. It is expected that trains will travel over this section of the track at a speed of 45 kilometres per hour. Find

   (a) the force exerted by the outer rail on the flanges of the wheels if the track is level and the mass of the train is 35 tonnes.

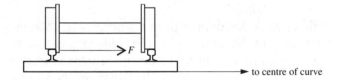

   (b) the height at which the outer rail should be raised above the inner rail to ensure that there is no pressure on the wheel flanges at the expected speed, given that the gauge of the track is 1.5 m (i.e. the rails are 1.5 m apart).

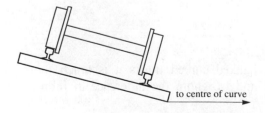

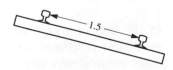

(a) The force acting on the train towards the centre of the curve is provided by the inward pressure on the wheels of the outer rail. The inner rail does not exert any force on the wheels.

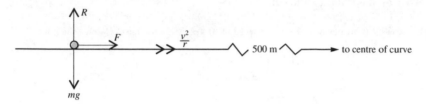

As we are not interested in any vertical forces, we will not resolve vertically.

$$45 \text{ km h}^{-1} = \frac{45\,000}{60 \times 60} \text{ m s}^{-1} = 12.5 \text{ m s}^{-1}$$

Newton's Law → $\quad F = 35\,000 \times \dfrac{(12.5)^2}{500} = 10\,937.5$

The force from the outer rail is 11 kN (2 sf)

(b) Each rail exerts a normal reaction on the train and these are probably different, but we will take R as the resultant normal reaction.

We will take m kg as the mass of the train, h metres as the difference in height of the two rails, and α as the angle at which the banked track is inclined to the horizontal.

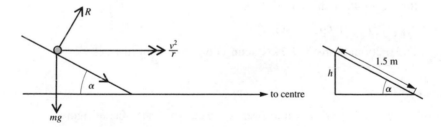

Resolving ↑ $\quad R \cos \alpha - 9.8m = 0$

Newton's Law → $\quad R \sin \alpha = \dfrac{m(12.5)^2}{500}$

Hence $\quad \dfrac{R \sin \alpha}{R \cos \alpha} = \dfrac{(12.5)^2 m}{500 \times 9.8m} \quad \Rightarrow \quad \tan \alpha = 0.03188\ldots$

∴ $\quad \alpha = 1.8° \quad (2 \text{ sf})$

Then $\quad h = 1.5 \sin \alpha = 0.0478\ldots$

The outer rail should be 48 mm above the inner rail (2 sf).

3. If the car in Example 1 is of mass 840 kg and drives round the curved section of the track at $36\,\mathrm{m\,s^{-1}}$, find the magnitude and direction of the lateral frictional force exerted by the track on the car.

(a) $36\,\mathrm{m\,s^{-1}}$ is greater than the design speed of $31.4\,\mathrm{m\,s^{-1}}$ so friction acts down the track.

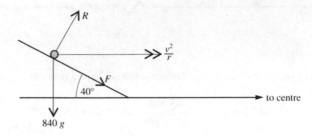

Resolving ↑ $R \cos 40° - F \sin 40° - 840g = 0$

i.e. $R \cos 40° - F \sin 40° = 8232$ [1]

Newton's Law → $R \sin 40° + F \cos 40° = 840 \times \dfrac{36^2}{120}$

                                                                      $= 9072$ [2]

[2] × cos 40° − [1] × sin 40°   gives

$$F \cos^2 40° + F \sin^2 40° = 9072 \cos 40° - 8232 \sin 40°$$

∴ $F(\cos^2 40° + \sin^2 40°) = 1658$

But $\cos^2 40° + \sin^2 40° = 1$

∴ $F = 1700$ (2 sf)

The frictional force is 1.7 kN acting down the banked track.

EXERCISE 2f

In this exercise give numerical answers corrected to 2 significant figures.

1. A train of mass 50 tonnes travels at $18\,\mathrm{m\,s^{-1}}$ round a bend which is an arc of a circle of radius 1.5 km. The track is horizontal.

(a)   Find the force exerted on the side of a rail.

(b)   On which rail does this act?

2. A locomotive is travelling at $80\,\mathrm{km\,h^{-1}}$, on a horizontal track, round a bend which is an arc of a circle. The locomotive has a mass of 10 tonnes. The lateral force exerted on a rail by the wheel flanges is 8000 N.

(a)   On which rail is this force acting?

(b)   Find the radius of the bend.

3. A road banked at $10°$ goes round a bend of radius $70\,\text{m}$. At what speed can a car travel round the bend without tending to side-slip?

4. On a level section of a race track a car can just go round a bend of radius $80\,\text{m}$ at a speed of $20\,\text{m}\,\text{s}^{-1}$ without skidding.

   (a) Find the coefficient of friction.

   On a section of the track that is banked at an angle θ to the horizontal a speed of $30\,\text{m}\,\text{s}^{-1}$ can just be reached without skidding, the coefficient of friction being the same in both cases. Taking the value of g as 10,

   (b) show that $\dfrac{\cos\theta + 2\sin\theta}{2\cos\theta - \sin\theta} = \dfrac{9}{8}$

   (c) find θ to the nearest degree.

5. A circular race track is banked at $45°$ and has a radius of $200\,\text{m}$. At what speed does a car have no tendency to side-slip? If the coefficient of friction between the wheels and the track is $\frac{1}{2}$, find the maximum speed at which the car can travel round the track without skidding.

6. An engine of mass $80\,000\,\text{kg}$ travels at $40\,\text{km}\,\text{h}^{-1}$ round a bend of radius $1200\,\text{m}$. If the track is level, calculate the lateral thrust on the outer rail. At what height above the inner rail should the outer rail be raised to eliminate lateral thrust at this speed if the distance between the rails is $1.4\,\text{m}$?

7. A race track has a circular bend of radius $50\,\text{m}$ and is banked at $40°$ to the horizontal. If the coefficient of friction between the car wheels and the track is $\frac{3}{5}$, find within what speed limits a car can travel round the bend without slipping either inwards or outwards.

8. A bend of a race track is banked at $45°$. The coefficient of friction between the wheels of a car and this track is $\frac{1}{2}$. The maximum speed at which the car can go round the bend without skidding is V.

   (a) Find the radius of bend in terms of V.

   (b) Find the design speed in terms of V.

9. The 'wall of death' at a fairground is in the form of the curved surface of a cylinder, of internal diameter $8\,\text{m}$, with its axis vertical. A motorcyclist rides round the wall on a path which is a horizontal circle. The coefficient of friction between wall and tyres is 0.9. Find the minimum speed he must maintain to stay on his circular path without slipping down the wall. State any assumptions you make in modelling this problem.

*10. The force which keeps an aircraft in the air is lift produced by the flow of air over the wings. The direction of this force is perpendicular to a line joining the wing-tips.

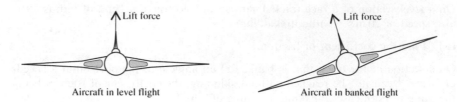

Aircraft in level flight   Aircraft in banked flight

(a) Explain why the aircraft must be banked to make it fly on a horizontal circular path.

(b) At what angle must it be banked in order to make it fly on such a path, of radius 2000 m, at $150\,\mathrm{m\,s^{-1}}$.

State any assumptions you make in modelling this problem.

*11. A bend of a race track, of radius r, is banked at $45°$. The coefficient of friction between the wheels of a car and the track is μ. The maximum speed at which the car can go round the bend without skidding is V_1 and the minimum speed at which it can go round without skidding is V_2.

(a) Show that $\dfrac{V_1{}^2}{rg} = \dfrac{1+\mu}{1-\mu}$.

(b) Obtain a similar expression involving V_2.

(c) Use your results from (a) and (b) to show that $\dfrac{V_1{}^2 V_2{}^2}{r^2 g^2} = 1$ and hence find r in terms of V_1 and V_2.

(d) Find the design speed in terms of V_1 and V_2.

CHAPTER 3

WORK AND ENERGY

BASIC FACTS

In Module E we saw that

- when a force moves the object to which it is applied, that force does work; the amount of work done is the product of the force and the distance through which the object moves in the direction of the force,

- if an object has the ability to do some work, the object possesses mechanical energy,

- when work is done to a system, the system gains an equivalent amount of mechanical energy; when a system does some work the system loses the equivalent amount of energy,

- if no work is done to, or by, a system, and there are no sudden changes in the motion, then the total amount of mechanical energy in the system remains constant. This is known as the Principle of Conservation of Mechanical Energy,

- the rate at which work is done is called power.

The two types of mechanical energy considered at that stage were:

Kinetic energy (KE) – the ability of an object to do work due to motion. The magnitude of the KE is given by $\frac{1}{2}mv^2$.

Gravitational potential energy (PE) – the ability to do work because of position relative to a chosen level. Its magnitude is given by mgh where h is the distance of an object *above* that chosen level.

The following exercise provides an opportunity to revise this topic.

EXERCISE 3a

1. A particle of mass 6.5 kg is moved from A to B under the action of the forces shown, where $\tan \theta = \frac{5}{12}$ and $AB = 20$ m.

   (a) Find the work done by each force shown in the diagram.

   (b) Find the total work done.

   (c) Find the change in kinetic energy.

   (d) If the speed at A is 5 m s^{-1}, find the speed at B.

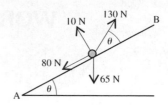

2. An escalator connects two floors of a building which are 6 m apart. It rises at 35° to the horizontal. Find the work done in taking up a man of mass 70 kg.

3. A crane lifts a load of mass 400 kg at a constant speed of 1.2 m s^{-1} for 30 s. Find the gain in potential energy.

4. Two boys are kicking a ball about. It has a mass of 0.5 kg. The ball comes towards one boy at 8 m s^{-1} and he passes it back at 12 m s^{-1}.

   (a) Find the work he does in bringing the ball to instantaneous rest.

   (b) Find the work he does in giving it a speed of 12 m s^{-1}.

5. A boy makes a slide on level icy ground. The slide is 6 m long. He runs up to the slide and steps onto it at a speed of 5 m s^{-1}. He reaches the other end at a speed of 4 m s^{-1}. His mass is 45 kg.

   (a) Find the loss of kinetic energy.

   (b) Find the resistance to motion.

   (c) Assuming that air resistance is negligible, find the coefficient of friction.

6. A particle slides down a smooth plane on a line AB, which is inclined at 40° to the horizontal. At A its velocity down the plane is 3 m s^{-1}. The height of A above B is 16 m. Find its velocity at B.

7. A car of mass 800 kg is travelling at 48 km h^{-1} when the driver sees a hazard ahead. He starts to brake at a distance of 25 m from the hazard. If the road is on horizontal ground, find the average braking force needed to stop the car in this distance.

8. Assuming that the car described in question 7 just manages to stop before the hazard, find

   (a) the time for which the brakes are applied

   (b) the average power exerted by the braking force.

9. Sue hopes to swing across the stream on a rope, which is attached to an overhanging tree at point A. The bank on the opposite side of the stream is 1.2 m higher than the bank on which she is standing. At what speed must she push off in order just to get there? State all the assumptions you make in modelling this problem.

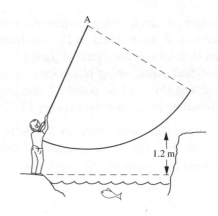

10. Peter Pan 'flies' across the stage on a harness, which slides along a smooth wire AB. End A of the wire is fixed at a height of 6 m and end B at a height of 5.5 m. The lowest point of the wire as he crosses the stage is at a height of 4 m. He starts from rest at A.

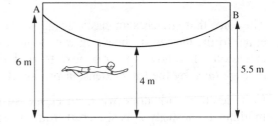

   (a) Find his maximum speed.

   (b) Find his speed at B.

11. A particle of mass m kg is pushed up a plane inclined at an angle θ to the horizontal. The coefficient of friction between the particle and the plane is μ. The particle has an initial speed u m s^{-1}. It travels a distance d m up a line of greatest slope of the plane and the speed then is v m s^{-1}.
   Find the work done

   (a) against the frictional force

   (b) against gravity

   (c) the total work done.

ELASTIC POTENTIAL ENERGY

An object possesses potential energy when the position of the object is such that releasing it from that position results in motion. We are familiar with the potential energy of an object which would fall from rest when released, but there is another type of potential energy.

Consider an elastic string fixed at one end and with a particle P attached to the other end. A force acting on P, away from the fixed end, stretches the string, and when that force is removed the particle begins to move, i.e. a particle attached to a stretched elastic string possesses potential energy because of its position. To distinguish this type of potential energy from gravitational potential energy, it is known as *elastic potential energy* (EPE).

We know that work done to a system causes an equivalent increase in the mechanical energy of the system, so the EPE of a stretched elastic string can be found by calculating the work done by the stretching force.

The Work Done in Stretching an Elastic String

The force that stretches an elastic string is not constant because it is at all times equal to the tension in the string, which, in turn, is directly proportional to the extension. It follows that one way to find the work done is to multiply the *average force* by the total extension produced.

Consider the work done when an elastic string, with a natural length a and modulus of elasticity λ, is stretched from an extension x_1 to an extension x_2.

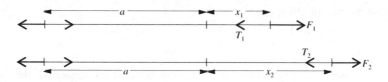

When the extension is x_1 $\qquad F_1 = T_1 = \dfrac{\lambda x_1}{a}$

When the extension is x_2 $\qquad F_2 = T_2 = \dfrac{\lambda x_2}{a}$

Over the period while the extension increases from x_1 to x_2,

the average extending force is $\frac{1}{2}(T_1 + T_2)$,

therefore the work done is given by $\frac{1}{2}(T_1 + T_2)(x_2 - x_1)$

i.e. **work done in stretching an elastic string**
= average tension × increase in extension

If x_1 is zero, i.e. the string is initially unstretched, then T_1 also is zero and the expression above can be simplified to give $\frac{1}{2}Tx$ where T is the final tension.

Further, using $T = \dfrac{\lambda x}{a}$, the work done can be expressed as $\dfrac{\lambda x^2}{2a}$

> **the amount of work needed to stretch an elastic string**
> **by an extension x is given by** $\dfrac{\lambda x^2}{2a}$

An alternative way to find the work done in stretching an elastic string, uses calculus.

When the extension is s, say, the extending force is $\dfrac{\lambda s}{a}$

Now if the string is further stretched by a small amount δs, the work required is given approximately by $\dfrac{\lambda s}{a} (\delta s)$.

The total work done in stretching the string from its natural length a to a length $(a + s)$ is therefore given approximately by $\displaystyle\sum_{0}^{x} \dfrac{\lambda s}{a} \delta s$.

Then, as $\delta s \to 0$, $\displaystyle\sum_{0}^{x} \dfrac{\lambda s}{a} \delta s \;\to\; \dfrac{\lambda}{a} \int_{0}^{x} s \, ds = \dfrac{\lambda x^2}{2a}$

The expressions derived above apply equally well to an extended *or compressed* elastic spring.

Examples 3b

1. The natural length of a light elastic string is 1.2 m and its modulus of elasticity is 18 N. Initially the string is just taut and is then stretched until it is 2 m long. Find the work done during the extension.

   The initial extension and the initial tension are both zero.

   The work done is given by $\dfrac{\lambda x^2}{2a}$, i.e. $\dfrac{18 \times 0.8^2}{2 \times 1.2}$

   The work done is 4.8 J.

2. The string described in example 1 is being held at a stretched length of 1.6 m when a force begins to stretch it further. The force acts until the work it has done amounts to 3 J.

    (a)   Find the final extension.

    (b)   State an assumption that has been made.

    (a)   The initial extension is 1.6 m; let the final extension be x m.

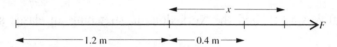

When extension is 0.4 m, the tension is $\dfrac{18 \times 0.4}{1.2}$ N,   i.e. 6 N

The final tension is given by $\dfrac{18x}{1.2}$

The average tension is given by $\frac{1}{2}(15x + 6)$

Work done = average tension × increase in extension

$$= \tfrac{1}{2}(15x + 6)(x - 0.4)$$

$\therefore \qquad 3 = \tfrac{1}{2}(15x^2 - 2.4)$

$\therefore \qquad 15x^2 = 8.4 \quad \Rightarrow \quad x = 0.7483\ldots$

The final extension is 0.748 m   (3 sf).

    (b)   We have assumed that the string has not exceeded its elastic limit.

3. One end of an elastic spring is fixed to a point A on a horizontal plane. The modulus of elasticity of the spring is λ, the natural length is a and the spring is strong enough to stand vertically. A particle of mass m is attached to the other end of the spring which is held at a distance a vertically above A. If the particle is allowed to descend gently to its equilibrium position find how much work is done in compressing the spring.

When the particle is in equilibrium

$$T = mg$$

Using Hooke's Law gives

$$T = \frac{\lambda x}{a}$$

$\therefore \qquad mg = \dfrac{\lambda x}{a} \quad \Rightarrow \quad x = \dfrac{mga}{\lambda}$

The work done in compressing the spring

is given by $\dfrac{\lambda x^2}{2a}$ i.e. $\dfrac{\lambda}{2a}\left(\dfrac{mg\,a}{\lambda}\right)^2$

The work done is $\dfrac{m^2 g^2 a}{2\lambda}$.

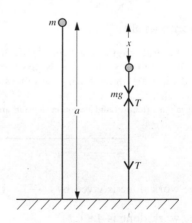

EXERCISE 3b

1. The natural length of an elastic string is 0.3 m and its modulus of elasticity is 8 N. Find the work done in

    (a) giving it an extension of 0.2 m

    (b) stretching it to a length of 0.6 m

    (c) stretching it from a length of 0.4 m, to a length of 0.7 m.

2. The natural length of a spring is 5*a* and its modulus of elasticity is 2*mg*. Find the work done in

    (a) compressing it to a length 4*a*

    (b) compressing it from length 3*a* to length 2*a*

    (c) compressing it until the thrust is *mg*.

3. The modulus of elasticity of a spring is 20 N and its natural length is 1.5 m. Find the work done in

    (a) stretching it until the tension is 8 N

    (b) stretching it to increase the tension from 8 N to 12 N.

4. An elastic string has a natural length of 2 m. The work done in extending it by 0.5 m is 6 J.

    (a) Find the modulus of elasticity.

    (b) Find the work done in stretching from its natural length to a length of 3 m.

5. Find the work done in stretching a rubber band round a roll of papers of radius 4 cm if the band when unstretched will just go round a cylinder of radius 2 cm and its modulus of elasticity is 0.5 N. Assume that the rubber band obeys Hooke's law.

6. A light elastic string is fixed at one end to a point A and a particle P attached to the other end hangs in equilibrium at a point B. The natural length of the string is 1.4 m, the modulus of elasticity is 8 N and the mass of the particle is 2 kg.

    (a) Find the depth of B below A.

    (b) Find the work done in stretching the string from its natural length to the length AB.

7. An upper body exerciser consists of a spring of length 20 cm with a handle attached at each end. The modulus of elasticity of the spring is 600 N. A man can stretch the spring to 25 cm in length.

    (a) Find the work he does in performing 15 repetitions of this exercise.

    (b) If these repetitions take him 20 seconds, find his power.

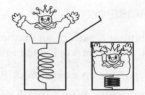

8. A jack-in-the-box toy comprises a box of
    depth 25 cm, with a spring fixed to its base.
    'Jack' is a puppet attached to the top of the
    spring. The natural length of the spring is
    30 cm. The height of the puppet is 10 cm
    and its mass is 0.2 kg.

    (a) When 'Jack' is in equilibrium with the box open the length of the spring is
         25 cm. Find the modulus of elasticity.

    (b) Making the assumption that the puppet does not compress or distort, find
         the work done in compressing the spring so that the box can be closed.

*9. A string of natural length $2a$ and modulus of elasticity λ has its ends attached to
     fixed points A and B, where   AB $= 3a$.   Find the work done when the mid-
     point C of the string is pulled away from the line AB to a position where triangle
     ABC is equilateral.

*10. The 'Beefee' exerciser is made of two bent rods ACD and BCE joined by a hinge
     at C. There are handles at A and B. A spring of natural length 15 cm connects
     D and E. The angles ACD and BCE are each 135° and the lengths AC, BC, CD
     and CE are each 40 cm. Gareth exercises by pressing together the handles until
     angle ACG is zero. He has to exert a force of 40 N on each handle to hold this
     position. Stating the assumptions which you are making, find

     (a) the modulus of elasticity of the spring.

     (b) the work done in one performance of this exercise.

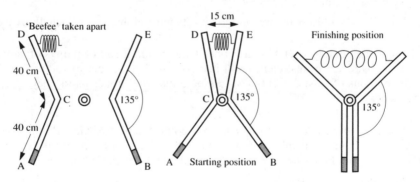

ENERGY PROBLEMS

We know that if no external work is being done to or by a system, the total
amount of mechanical energy remains constant, and problems involving the
conservation of kinetic energy and gravitational potential energy have already
been considered.

Now that the elastic potential energy of a stretched elastic string can be found (the work done in stretching an elastic string from its natural length is equal to the elastic potential energy in the string) problems can be tackled where three types of mechanical energy, KE, PE and EPE, arise. As there are now likely to be changes in all three types, equating the *total* mechanical energy in one position with the total in another position is wiser than trying to juggle with which types of energy are decreasing and which increasing.

Examples 3c

1. A particle of mass 2 kg is attached to one end of an elastic string of length 1 m and modulus of elasticity 8 N which is lying on a smooth horizontal plane. The other end of the string is fixed to a point A on the plane and when the string is just taut the particle is at a point B. The particle is pulled away from A until it reaches the point C where   AC = 1.5 m   and is held in that position.

   (a) Find the elastic potential energy in the string.

   The particle is then released from rest.

   (b) Find the velocity of the particle when it passes through the point B.

   (c) What is its velocity when it passes through A?

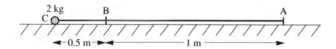

(a) When the particle is at C,

   the EPE is given by $\dfrac{\lambda x^2}{2a}$,   i.e. $\dfrac{8 \times (0.5)^2}{2 \times 1}$

   The EPE in the string is 1 J.

(b) In moving from C to B, there is no change in PE as the motion takes place on a horizontal plane.

At C,   KE = 0   and   EPE = 1 J

At B,   KE = $\frac{1}{2}mv^2$ = $\left(\frac{1}{2}\right)(2)v^2$ and   EPE = 0

Conservation of mechanical energy gives

$$0 + 1 = v^2 + 0 \quad \Rightarrow \quad v = 1$$

The velocity of the particle at B is 1 m s^{-1} towards A.

(c)   When the particle reaches B the string is no longer stretched so there is no EPE in the string.

Between B and A there is no change in EPE therefore the KE remains constant.

∴   the velocity at A is $1\,\mathrm{m\,s^{-1}}$ in the direction CA.

Note that the speed of the particle remains constant until it reaches a point D on the opposite side of A where   AD = 1 m.   Beyond this point the string again begins to stretch and contains EPE.

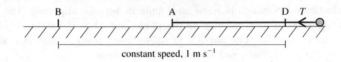

constant speed, $1\ \mathrm{m\ s^{-1}}$

2. **A light elastic spring of length 1 m and modulus of elasticity 7 N has one end fixed to a point A. A particle of mass 0.5 kg hangs in equilibrium at a point C vertically below A.**

(a)   **Find the distance AC.**

The particle is raised to the point B, between A and C, where   AB = 1 m, and is released from rest. Find

(b)   **the speed of the particle as it passes through C**

(c)   **the distance below B of the point D where the particle first comes to rest.**

(a)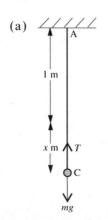

At C $T = mg$

i.e. $T = 0.5 \times 9.8 = 4.9$

Using Hooke's Law gives $T = \dfrac{7x}{1}$

∴ $7x = 4.9$ ⇒ $x = 0.7$

The extension is 0.7 m.

∴   The distance AC is 1.7 m.

(b)

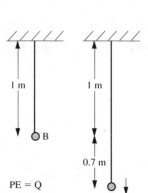

We will use conservation of ME from B to C.

At B $PE = mgh = 0.5 \times 9.8 \times 0.7 = 3.43$

$KE = 0$

$EPE = 0$

At C $PE = 0$

$KE = \frac{1}{2}mv^2 = 0.25v^2$

EPE in the string is given by $\dfrac{\lambda x^2}{2a}$

i.e. $EPE = \dfrac{7 \times 0.7^2}{2 \times 1} = 1.715$

Conservation of ME gives

$$3.43 + 0 + 0 = 0 + 0.25v^2 + 1.715$$

$\therefore$ $v^2 = 6.86$

$\Rightarrow$ $v = 2.619\ldots$

The speed at C is $2.62\,\mathrm{m\,s^{-1}}$ (3 sf).

(c)

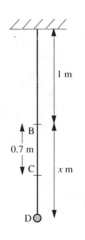

We could use conservation of ME from C to D but that involves the speed at C which *might* not be correct. By working from B to D we avoid this.

At B $PE = (0.5)(9.8)(x)$

$KE = 0$ and $EPE = 0$

At D $PE = 0$ and $KE = 0$

$$EPE = \frac{7x^2}{2 \times 1}$$

Conservation of ME gives

$$4.9x = 3.5x^2 \qquad [1]$$

$\therefore$ $x = 0$ or 1.4

When $x = 0$ the particle is at B so, at D, $x = 1.4$.

$\therefore$ D is 1.4 m below C.

Note that the two positions of instantaneous rest, i.e. B and D, are at equal distances from the equilibrium position C.

3. One end of a light elastic string, with natural length 0.8 m and modulus of elasticity 16 N, is fixed at a point A on a smooth plane inclined at 45° to the horizontal. A particle of mass 1 kg, attached to the other end of the string, rests in equilibrium at a point E on the plane.

(a) Find the distance AE.

The particle is pulled down the plane to a point C, where   AC = 1.6 m,   and is then released from rest

(b) Find the speed of the particle as it passes through E.

(a)

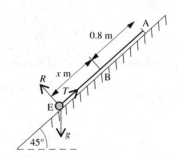

Considering the forces acting on the particle along the plane gives

$$T - g \cos 45° = 0 \qquad \Rightarrow \qquad T = 6.929\ldots$$

Using Hooke's Law, $T = \dfrac{\lambda x}{a}$,   gives

$$6.929\ldots = \frac{16x}{0.8} \qquad \Rightarrow \qquad x = 0.346\ldots$$

The length of AE is 1.15 m   (3 sf).

(b)

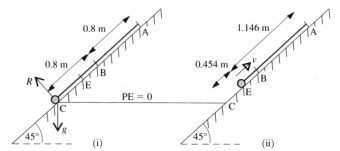

The normal contact force on the particle is perpendicular to the direction of motion, and the plane is smooth, so no work is done.

When the particle is at C, total mechanical energy $=$ EPE

$$= \frac{16(0.8)^2}{2 \times 0.8} J = 6.4 J$$

When the particle is at E, total mechanical energy $=$ EPE $+$ PE $+$ KE

$$= \left\{ \frac{16(0.346)^2}{2 \times 0.8} + 1 \times g \times 0.454 \sin 45° + \tfrac{1}{2} \times 1 \times v^2 \right\} J$$

$$= (0.5v^2 + 4.343 \ldots) J$$

Using conservation of mechanical energy between (i) and (ii) gives

$$0.5v^2 + 4.343 = 6.4$$

$\Rightarrow \qquad\qquad v = 2.028 \ldots$

The velocity of the particle at E is $2.03\,\text{m s}^{-1}$ (3 sf).

4. In a fairground 'test your strength' machine, a spring is fixed at one end and lies, just taut, in a horizontal groove. A metal cylinder is attached to the other end.

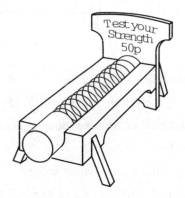

The would-be strong men strike the cylinder so as to compress the spring, and the machine records the compression achieved when the cylinder first comes to rest. The natural length of the spring is 1 m, the modulus of elasticity is 60 N and the mass of the cylinder is 4 kg. If a competitor gave the cylinder an initial speed of $3\,\text{m s}^{-1}$, find the recorded compression if

(a) the groove is smooth

(b) the coefficient of friction between cylinder and groove is 0.2.

State any assumptions you have made in your solution, with comments on their validity.

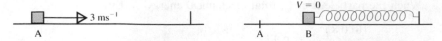

At A the only mechanical energy is KE

so total energy at A is $\frac{1}{2}(4)(3^2)$ J, i.e. 18 J

At B the only mechanical energy is EPE

so total energy at B is $\dfrac{60x^2}{(2)(1)}$ J, i.e. $30x^2$ J

These expressions apply to both parts of the question.

(a) No work is done to the system once the cylinder is set moving so we can use conservation of mechanical energy, giving

$$30x^2 = 18 \quad \Rightarrow \quad x = 0.77\ldots$$

The recorded compression is 0.77 m (2 sf).

(b) If the groove is rough, work is done by the frictional force, so energy is not conserved. The frictional force opposes the motion of the cylinder so the work it does causes a loss of energy in the system.

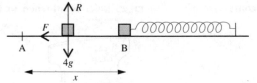

Resolving vertically for the cylinder gives $\qquad R = 4g$

The frictional force, F, is given by μR i.e. $\qquad F = 0.2 \times 4g = 7.84$

Friction acts for a distance x. So work done by friction is Fx, i.e. $7.84x$

$$\text{work done} = \text{loss in energy}$$

$$= \text{initial energy} - \text{final energy}$$

i.e. $\qquad 7.84x = 18 - 30x^2 \quad \Rightarrow \quad 30x^2 + 7.84x - 18 = 0$

Solving this quadratic equation by using the formula gives

$$60x = -7.84 \pm \sqrt{(7.84^2 + 2160)} \quad \Rightarrow \quad x = 0.654\ldots$$

The recorded compression is 0.65 m (2 sf).

(c) Assumptions made are:
the cylinder is modelled as a particle; the spring is light

and for part (b)
the groove is uniformly rough; friction between the *spring* and the groove can be neglected.

EXERCISE 3c

In questions 1 and 2 a particle of mass 2 kg is attached to one end of a light elastic string of natural length a metres and modulus of elasticity λ newtons. The other end of the string is attached to a fixed point A on a smooth horizontal surface.

1. $\lambda = 5$, $a = 0.8$, $AB = 1\,m$ and $BC = 1\,m$.

The particle is held at C, with the string stretched and released from rest.

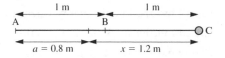

(a) Find the elastic potential energy when the particle is at C.

(b) Find the elastic potential energy when the particle is at B.

(c) Find the kinetic energy when the particle is at B.

(d) Find the speed of the particle as it passes through B.

2. $\lambda = 8$, $a = 0.5$, $AB = 0.5\,m$ and $BC = 0.7\,m$.

The particle is projected from point B with velocity $10\,m\,s^{-1}$ in the direction BC.

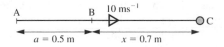

(a) Find the elastic potential energy when the particle is at B.

(b) Find the kinetic energy when the particle is at B.

(c) Find the elastic potential energy when the particle is at C.

(d) Find the speed of the particle as it passes through C.

In questions 3 and 4 a particle of mass 3 kg is attached to one end of a light elastic spring of natural length a metres and modulus of elasticity λ newtons. The other end of the spring is attached to a fixed point A on a horizontal surface. The coefficient of friction between the particle and the surface is μ.

3. $\mu = 0$, $\lambda = 800$, $a = 2$, $AC = 2\,m$.

The particle is projected from C with speed $20\,m\,s^{-1}$ and it has a speed of $10\,m\,s^{-1}$ as it passes through B.

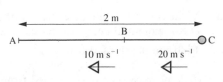

(a) Find the elastic potential energy when the particle is at C.

(b) Find the kinetic energy when it is at C.

(c) Find the kinetic energy when it is at B.

(d) Find the elastic potential energy when it is at B.

(e) Find the distance AB.

4. $\lambda = 150$, $a = 3$,   AB $= 2$ m   and   BC $= 3$ m.

The particle is released from rest at C
and it has a speed of $4\,\mathrm{m\,s^{-1}}$ as it passes
through B.

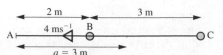

(a)   Find the elastic potential energy
      when the particle is at C.

(b)   Find the elastic potential energy when it is at B.

(c)   Find the kinetic energy when it is at B.

(d)   Find the work done by friction during the motion from C to B.

(e)   Find the value of μ.

5. A particle of mass $2\,\mathrm{kg}$ is suspended from a point A by a light elastic spring of
natural length $1\,\mathrm{m}$ and modulus of elasticity $80\,\mathrm{N}$. The particle is initially held at
a point B, which is $0.6\,\mathrm{m}$ vertically below A, with the spring compressed. It is
then released from rest at B. In the subsequent motion the particle is at point C
when the spring has reached its natural length. Take the gravitational potential
energy to be zero at point C.

(a)   When the particle is at B find the values of
      (i)   the elastic potential energy   (ii)   the gravitational potential energy
      (iii)   the kinetic energy.

(b)   When the particle is at C find the value of the elastic potential energy.

(c)   Find the speed of the particle as it passes through C.

6. In this question use $g = 10\,\mathrm{m\,s^{-2}}$.

One end of a light elastic string of natural length $2\,\mathrm{m}$ and modulus of elasticity
$120\,\mathrm{N}$ is fixed at a point A; the other end carries a particle of mass $1\,\mathrm{kg}$. The
particle is released from rest at the point A and drops vertically. It first comes to
rest at a point B vertically below A and the string then has an extension of x metres.

(a)   Write down the kinetic energy of the particle when it is at A and find, in
      terms of x,
      (i)   the elastic potential energy
      (ii)   the gravitational potential energy relative to B.

(b)   Write down the kinetic energy of the particle when it is at B and find the
      elastic potential energy in terms of x.

(c)   Find the depth of B below A.

7. A particle of mass $5\,\mathrm{kg}$ is suspended from a point A by a light spring of
natural length $0.4\,\mathrm{m}$. When in equilibrium the particle is at a point B, vertically
below A, where   AB $= 1.8\,\mathrm{m}$. The particle is pulled down to point C, also
vertically below A, where   AC $= 3\,\mathrm{m}$. It is then released from rest at C.

(a)   Find the modulus of elasticity.

(b)   Find the speed of the particle as it passes through B.

8. A light spring, of natural length $4a$ and modulus of elasticity $2mg$, is fixed at a point A and a particle of mass m is attached to the other end. When the particle is suspended in equilibrium, it is at a point B vertically below A. It is projected vertically downwards from B with speed v and first comes to rest at a point C where $AC = 8a$.

(a) Find the distance AB.

(b) Show that $v = \sqrt{2ga}$.

9. A particle of mass 2 kg lies on a smooth plane which is inclined at an angle of 30° to the horizontal. The particle is attached to one end of a light elastic string of natural length 1 m and modulus of elasticity 20 N. The other end of the string is attached to a point A on the plane. The particle can rest in equilibrium at a point B on the plane.

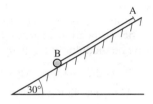

(a) Find the extension of the string when the particle is at B.

If the particle is released from rest at A,

(b) find the speed with which it will pass through B.

(c) find the distance it has travelled down the plane from A when it first comes to rest.

*10. In a pin-ball machine the ball, of mass 30 g, is propelled from a cup, of mass 90 g, which is attached to a spring. The spring has modulus of elasticity 9 N and natural length 15 cm. The mechanism allows the spring to be given a compression of 10 cm. It is then released and the ball and cup move over the smooth horizontal surface of the table. The ball leaves the cup when the spring reaches its natural length.

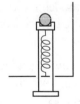

(a) Find the elastic potential energy of the spring when it is fully compressed.

(b) Find the kinetic energy of the cup and ball combined, at the moment when the spring reaches its natural length.

(c) Find the speed with which the ball leaves the cup.

(d) Find the kinetic energy of the cup at the moment when the ball has just left it.

(e) Find the extended length of the spring at the moment when the cup first comes to rest.

*11. An aircraft of mass 5000 kg lands on the deck of an
aircraft-carrier at a speed of $50\,\mathrm{m\,s^{-1}}$ relative to the
deck. It catches on the mid-point of an elastic cable
which lies, just taut, across the deck at 90° to its
path. The cable has a natural length of 60 m. The
plane is brought to rest in a distance of 40 m by the
stretching cable and by a retarding force of
magnitude 8000 N in the direction opposite to its
motion, which is produced by its engines. Find the
modulus of elasticity of the cable.

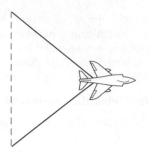

*12. A and B are two fixed points on a smooth horizontal plane. AB = 3.3 m. A spring,
of natural length 0.9 m and modulus of elasticity 6 N, has one end attached to point A
and the other end to a particle P, of mass 0.03 kg. A second spring, of natural length
1.5 m and modulus of elasticity 8 N has its ends attached to the particle and to point B.

(a)   Find the distance AP when the particle is in equilibrium.

(b)   The particle is released from rest at point C where   AC = 0.5 m.
  (i)   Find its velocity as it passes through the equilibrium position.
  (ii)   Find the length of AP when the particle first comes to rest.

REFINING MATHEMATICAL MODELS

A mathematical model of a situation is produced by making certain assumptions.
These may include assuming that certain objects are small enough to be treated as
particles, other subjects are light enough to be treated as weightless, air resistance
can be ignored, relevant physical laws are obeyed, and so on.

The model is then used to predict results. If these results, compared with those
observed in trials, are not accurate enough to be useful, then the model requires
improvement.

One way to achieve this is to include one or more of the quantities previously
ignored. Which of these to choose depends upon assessing how reasonable the
original assumptions were.

Consider, for example, the loaded weighing machine given below, modelled by a
light spring that obeys Hooke's Law, with a particle placed on it as shown.

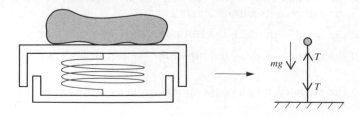

One of the assumptions made in forming this model is that the platform is light. If unreliable results are produced, a refined model could take into account the mass of the platform. Whether or not the refined model produces better estimates can only be assessed by comparing predicted results with observed results.

EXERCISE 3d

1. A climber of mass 60 kg is attached by a rope of length 30 m to a point on a vertical rock-face. When she is 20 m above the point to which the rope is attached she slips and falls. She first comes to rest after falling through a vertical distance of 57 m. Assuming that she is not retarded by contact with the rock-face during her fall, and that the mass of the rope can be neglected, state any further assumptions you would make in order to model this situation.

   (a) Use your model to find
      (i) the modulus of elasticity of the rope
      (ii) her speed at the moment that the rope becomes taut.
   (b) What features of the model might lead to unreliable estimates? State any adjustments that could be made.

*2. The Department of Transport in Ruritania, having invested in new trains for its railways, decides to replace all the old buffers in the stations. A technical team, looking into the design of new buffers, decides first of all to produce a mathematical model.

   The first model treats the train as a particle, of mass M kilograms, hitting the buffers at a speed of v metres per second and being brought to rest in a distance of d metres. The buffers are modelled as a spring, of natural length a metres, modulus of elasticity λ, fixed at one end and able to take a maximum compression of c metres. The mass of the buffers is taken to be negligible, being small compared to the mass of the train. Using this model there is no loss of energy when the particle collides with the spring. In this first model it is assumed that the retarding force produced by the brakes is negligible.

   (a) Assuming that the spring does not become fully compressed, obtain d in terms of M, v, a and λ.
   (b) Show that if the train is to be brought to rest before the spring is fully compressed then it is necessary to satisfy the condition $\lambda > \dfrac{M a v^2}{c^2}$.

   It is decided to refine the model by introducing a retarding force R, produced by the brakes.

   (c) Again assuming that the spring does not become fully compressed, show that now
   $$d = \frac{\sqrt{a^2 R^2 + \lambda M a v^2} - aR}{\lambda}$$
   (d) Show that this refined model includes the result obtained in (a) from the original model.

3. A bungee jumping event is to take place from a suspension bridge which is 50 m above a river. The jumpers dive from the bridge attached to an elastic rope. You are required to consider some suitable ropes for this activity. The aim is that the jumper should be brought to rest just above the surface of the water. Carry out the investigation for the case of a man of mass 70 kg, using a model which treats the man as a particle and the rope as light, and assume that the initial velocity is zero. State any other assumptions that you make.

Use the following variables:

natural length of the rope, a metres,

modulus of elasticity of the rope, λ newtons,

extension of the rope when the man reaches the water, x metres,

extension of the rope if the man is suspended in equilibrium on it, e metres,

the speed with which he passes through this equilibrium position, V metres per second (this is the greatest speed he reaches).

(a) Obtain an expression for λ in terms of a and x.

(b) Obtain an expression for e in terms of a and λ.

(c) Obtain an expression for V in terms of a, e and λ.

(d) Complete the following table of values.

a	10	20	30	40	45
x					
λ					
e					
$a+e$					
V					

(e) Can you suggest any factors which might make some of these lengths of rope unsuitable?

(f) Which assumptions might you consider changing in order to refine the model?

CONSOLIDATION A

SUMMARY

Elastic Strings and Springs

An elastic string is one that can be stretched and will return to its original length.

When it is stretched it is said to be taut.

Its unstretched length is its natural length and in this state it is said to be just taut.

An elastic string is stretched by two equal forces acting outwards, one at each end of the string, which are equal to the tension in the string.

These properties apply also to a stretched elastic spring but a spring, unlike a string, can be compressed. It is then said to be in thrust and exerts an outward push at each end, equal to the compressing force.

Hooke's Law states that the extension, x, in an elastic string or spring is proportional to the tension, T, in the string, i.e.

$$T = \frac{\lambda x}{a}$$

where a is the natural length of the string
and λ is its modulus of elasticity.

The unit for λ is the newton and the value of λ is equal to the force that doubles the length of the string.

A string that has reached its elastic limit no longer obeys Hooke's Law and will not return to its original length if stretched further.

Circular Motion

The angular velocity of a particle describing a circle is represented by $d\theta/dt$ or ω, and is measured in radians per second or revolutions per second.

A particle travelling in a circle of radius r metres, with a constant angular velocity $\omega \, \text{rad s}^{-1}$, has a speed round the circumference of $r\omega \, \text{m s}^{-1}$.

When a particle describes a circle of radius r at a constant speed v (or constant angular velocity ω):

- the acceleration is directed towards the centre of the circle and is of magnitude $r\omega^2$ or v^2/r,
- a force of magnitude $mr\omega^2$ or mv^2/r must act towards the centre.

A string fixed at one end and carrying at the other end a particle performing horizontal circles, is known as a conical pendulum.

The force that enables a vehicle to travel on a horizontal circular path can be provided by friction, or by pressure from rail flanges, or by banking the track on which it moves.

Elastic Potential Energy

A stretched elastic string or a stretched or compressed elastic spring possesses elastic potential energy (EPE) equal to the amount of work needed to provide the extension or compression.

The amount of EPE in an elastic string stretched by an extension x from its natural length a is given by

$$\frac{\lambda x^2}{2a}$$

If a string is already stretched by an amount x_1 and is then further stretched until the extension is x_2, the work done in producing the extra extension is given by either of the following expressions:

(average of initial and final tensions) $\times (x_2 - x_1)$

$$\frac{\lambda}{2a}(x_2{}^2 - x_1{}^2)$$

MISCELLANEOUS EXERCISE A

In questions 1 and 2 a problem is set and is followed by a number of suggested responses. Choose the correct response.

1. If the force needed to compress a spring to half of its natural length is T, then the force needed to stretch it to twice its natural length is

   A $\frac{1}{4}T$ B $\frac{1}{2}T$ C T D $2T$

2. If the work done in stretching a spring to twice its natural length is E, then the work done in compressing it to half of its natural length is

   A $\frac{1}{4}E$ B E C $4E$ D $16E$

In question 3 a problem is set and is followed by a number of statements. Decide whether each statement is true (T) or false (F).

3. The diagram shows an elastic string of natural length a and modulus of elasticity $2mg$, fixed at one end to a point A and with a particle P, of mass m, attached to the other end. P is released from rest at a point C where $AC = 2a$.

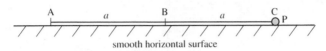

smooth horizontal surface

(i)   At C the acceleration is $2g$.
(ii)   At C the elastic potential energy is $2mga$.
(iii)   When P reaches B the kinetic energy is zero.
(iv)   When P reaches B the acceleration is zero.
(v)   When P reaches A the velocity is zero.

4.

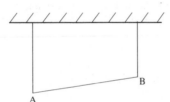

The diagram shows a non-uniform rod AB of length l hanging in equilibrium, suspended by two vertical elastic strings. The strings have the same unstretched length and the same modulus of elasticity. In the equilibrium position shown, the extension of the string attached to A is twice the extension of the string attached to B. Find the distance of the centre of gravity of the rod AB from A.

(UCLES)ₛ

5.

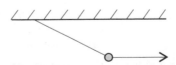

The diagram shows a light elastic string, of natural length a and modulus mg, fixed at one end to a point on the ceiling of a room. To the other end of the string is attached a particle of mass m. A horizontal force is applied to the particle and when the system is in equilibrium the length of the string is $3a$. When the system is in equilibrium

(a)   find the tension in the string,

(b)   show that the string makes an angle of $30°$ with the horizontal.   (ULEAC)

6. A particle P, of mass 1 kg, is connected to a light inextensible string of length 0.5 m. The other end of the string is tied to a fixed point O on a smooth horizontal plane. P moves on the plane in a horizontal circle, centre O, with uniform speed. Given that the string will break when the tension exceeds 8 N, show that P can rotate at 38 revolutions per minute without breaking the string.

(ULEAC)

7.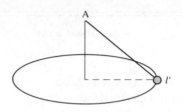

The conical pendulum, shown in the diagram, consists of a light inextensible string which has one end attached to a fixed point A. A particle P, of mass m, is attached to the other end of the string. The particle P moves with constant speed completing 2 orbits of its circular path every second and the tension in the string is $2mg$.

Find, to the nearest cm,

(a) the radius of the circular path of P,

(b) the length of the string.   (ULEAC)

8. In the sport of bungee jumping, a light elastic rope has one end attached to the participant and the other end attached to a fixed support on a high bridge. The participant then steps off the bridge and falls vertically. Given that a particular participant has mass 60 kg, and that the rope has natural length 30 m and modulus of elasticity 588 N, find, in metres to one decimal place, the distance of the participant below the point of support at the first instant of instantaneous rest.

(You may assume that the participant does not reach the ground.)   (AEB)ₛ

9. A light elastic string of natural length 0.3 m has one end fixed to a point on a ceiling. To the other end of the string is attached a particle of mass M. When the particle is hanging in equilibrium, the length of the string is 0.4 m.

(a) Determine, in terms of M and g, the modulus of elasticity of the string.

A horizontal force is applied to the particle so that it is held in equilibrium with the string making an angle α with the downward vertical. The length of the string is now 0.45 m.

(b) Find α, to the nearest degree.   (ULEAC)

10. A light elastic string, of natural length 2 m and modulus 39.2 N, has one end attached to a fixed point A. A particle P, of mass 1 kg, is attached to the other end of the string.

(a) Show that, when P hangs in equilibrium, the length of AP is $2\frac{1}{2}$ m.

The particle P is released from rest at A and falls vertically. Use the work–energy principle to calculate

(b) the speed, to 3 significant figures, of P at a distance $2\frac{1}{2}$ m below A,

(c) the greatest length of the string.   (ULEAC)

11. Two light strings, AB and BC, are each attached at B to a particle of mass m. The string AB is elastic, of natural length $2a$ and modulus $3mg$. The string BC is inextensible and of length $3a$. The ends A and C are fixed with C vertically below A and AB $= 5a$.

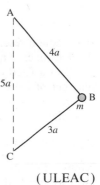

The particle moves with constant speed in a horizontal circle, with both strings taut and AB $= 4a$, as shown in the figure.

(a) Find the tension in the string AB.

(b) Find the tension in the string BC.

(c) Show that the speed of the particle is

$$\sqrt{\left(\frac{44}{5}\,ga\right)}$$

         (ULEAC)

12. A boy whirls a conker of mass 25 g in a horizontal circle at a constant angular speed. The length of the string supporting the conker is 40 cm. Assume that the boy's hand is at rest.

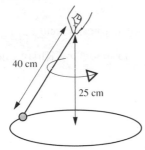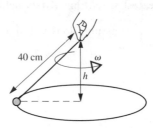

(a) Draw a force diagram showing the forces acting *on the conker*.

(b) Find the tension in the string when the conker is 25 cm below the point of support.

(c) Find the magnitude of the acceleration of the conker.

(d) Find the speed of the conker in m s^{-1}.

(e) The angular speed ω rad s^{-1} of the conker is changed and then becomes constant again. Derive an expression relating ω to the depth, h m, of the conker below the point of support.

(f) Explain why the conker can never be whirled in a horizontal circle level with the point of support.   (UODLE)$_s$

13.

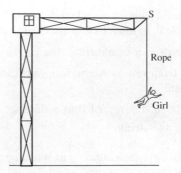

A mountain rescue team is investigating whether or not to use a new type of flexible rope. They take a length of 20 m of the rope. One end is attached to the top S of a fixed crane and a harness is attached to the other end. Kirsty, a member of the team, whose mass together with that of the harness is 60 kg, is lowered gently until she hangs at rest. The stretched rope is then 21 m long. By modelling the rope as a light elastic string and Kirsty as a particle,

(a)   estimate the modulus of elasticity of the rope.

Kirsty climbs back to S and releases herself to fall vertically, strapped in the harness attached to the rope. She comes to instantaneous rest at the point C at the end of her descent.

Estimate

(b)   the length SC of the stretched rope,

(c)   the greatest speed that Kirsty achieves during her descent.

State any further assumptions you have made in modelling Kirsty's descent from S to C.   (ULEAC)ₛ

14. A child of mass M kg sits on one of the seats of a 'rotating swing', and moves in a horizontal circle of radius 10 m with constant speed, completing one circuit every 5 s. Each seat has mass m kg. Find the angle between the single chain supporting the seat and the vertical.

Give a reason why the chains are all at the same angle to the vertical, irrespective of the mass of the occupant.

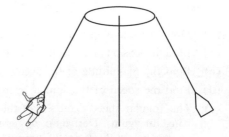

(UCLES)ₛ

15.

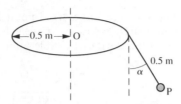

A light, inextensible string of length 0.5 m is attached to a point on the circumference of a disc of radius 0.5 m. To the other end of the string is attached a small stone P of mass M kilograms. The disc is made to rotate in a horizontal plane, about a vertical axis through its centre O. *At each instant, the string lies in a vertical plane through O.* Given that the disc rotates at a constant rate of 2 rad s^{-1}, and that the string makes a constant angle α with the downward vertical, as shown in the diagram, show that

$$\cot \alpha + \cos \alpha = 4.9.$$

A physical force is taken as negligible in this question and its omission is implied by the sentence in italics. Describe the force in words. $(\text{ULEAC})_s$

16.

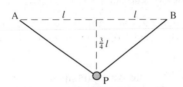

The diagram shows a particle P, of mass $6M$, suspended by two identical light elastic strings from the points A and B which are fixed and at a horizontal distance $2l$ apart. Each string has natural length l and P rests in equilibrium at a vertical distance $\frac{3}{4}l$ below the level of AB. Determine

(a)   the tension in either string,

(b)   the modulus of elasticity of either string. (ULEAC)

17. The figure shows two light inextensible strings AB and BC, each of length d, attached at B to a particle of mass m. The other ends A and C are fixed to points in a vertical line with A above C. With both strings taut and $\angle ABC = 120°$, the particle moves in a horizontal circle with constant angular speed ω. Find, in terms of m, g, d and ω,

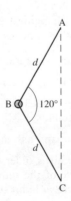

(a)   the tension in string AB,

(b)   the tension in string BC,

(c)   Show that $\omega^2 \geqslant \left(\dfrac{2g}{3d}\right) \sqrt{3}$.

(ULEAC)

18. A particle of mass 1.8 grams is attached to a fixed point A by a string 1 metre long and describes a horizontal circle below A. Given that the breaking tension of the string is 3 newtons, find the greatest possible number of revolutions per second.   (SMP)$_s$

19. In the dangerous sport of bungee diving an individual attaches one end of an elastic rope to a fixed point on a river bridge. He/she is attached to the other end and jumps over the bridge so as to fall vertically downwards towards the water. The rope should be such that the diver comes to rest just above the surface of the water. In order to find out which particular ropes are suitable experiments are carried out with weights attached to the rope rather than people. In one experiment it was found that when a weight of mass m was attached to a particular rope of natural length a and dropped from a bridge at a height of $3a$ above the water level then the weight just reached the level of the water. Show that the modulus of elasticity of the rope is $3mg/2$.

State, with justification, whether the above result is an underestimate or overestimate of the modulus of elasticity of the rope.

The weight of mass m is then removed and a weight of mass $5m/2$ is then attached to this rope and dropped from the same height so that the weight enters the water.

When the weight emerges from the water its speed has been reduced to zero by the resistance of the water. Show by using conservation of energy or otherwise and assuming that the rope does not slacken, that the subsequent speed v of the weight at height h above the water level is given by

$$v^2 = \frac{gh}{5a}(2a - 3h).$$ (WJEC)$_s$

20. A child of mass 30 kg keeps herself amused by swinging on a 5 m rope attached to an overhanging tree. She is holding on to the lower end of the rope and 'swinging' in a horizontal circle of radius 3 m.

(a) Draw a diagram to show the forces acting on the girl.

(b) Find the tension in the rope.

(c) Show that the time she takes to complete a circle is approximately 4 seconds.

(d) State any assumptions that you have made about the rope.

(e) The girl's older brother then swings, on his own, on the rope in a horizontal circle of the same radius. Show that the tension in the rope is now $5mg/4$ where m is his mass.

Find the time that it takes for him to complete one circle.   (AEB)

21. A string of natural length $2a$ and modulus of elasticity λ has its ends fixed to two points, A and B, which are at the same horizontal level and at a distance $2a$ apart. The centre of the string is pulled back to a point C in the same horizontal plane as A and B such that ABC forms an equilateral triangle.

Find the tension in the stretched string and the energy stored in it.

A small mass m is placed inside the stretched string at the point C and the string released. The mass is catapulted through the mid-point of AB.

Neglecting the effects of gravity find the speed of the mass as it leaves the string.
                                                                                      (MEI)$_p$

22. The buffer at the end of a railway siding is designed to stop trucks that run into it without any damage being done. The system in the buffer that absorbs the energy of the truck is modelled by an elastic spring which begins to be compressed when the truck runs into the buffer. Once the spring has been compressed, it is prevented from returning to its natural length. Tests show that a truck of mass 10 tonnes moving at $0.5\,\mathrm{m\,s^{-1}}$ is stopped by the buffer when the spring has been compressed by 0.4 m. The maximum compression allowed for in the design of the buffer is 1.2 m. Calculate the maximum compression force that the buffer system is able to withstand.
                                                                                      (UCLES)$_s$

23. Two light springs are joined and stretched between two fixed points A and C which are 2 m apart as shown in the diagram. The spring AB has natural length 0.5 m and modulus of elasticity 10 N. The spring BC has natural length 0.6 m and modulus of elasticity 6 N. The system is in equilibrium.

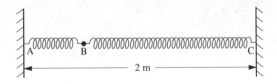

(i)   Explain why the tensions in the two springs are the same.
(ii) Find the distance AB and the tension in each spring.
(iii) How much work must be done to stretch the springs from their natural length to connect them as described above?

A small object of mass 0.012 kg is attached at B and is supported on a smooth horizontal table. A, B and C lie in a straight horizontal line and the mass is released from rest at the mid-point of AC.

(iv) What is the speed of the mass when it passes through the equilibrium position of the system?
                                                                                      (MEI)

In questions 24 to 27 a problem is set and is followed by a number of suggested responses. Choose the correct response.

24. A small sphere is moving round inside the smooth rim of a circular tray of radius 0.2 m. The mass of the sphere is 1 kg and its speed is $3\,\mathrm{m\,s^{-1}}$. The force exerted on the sphere by the rim of the tray is

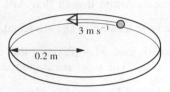

A   1.8 N   B   −45 N   C   −1.8 N   D   45 N

For questions 25 to 27 use this diagram of a conical pendulum. The particle P is travelling with constant speed v.

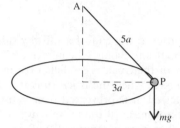

25. The tension in the string is

A $\frac{3}{5}mg$ B $\frac{4}{5}mg$ C $\frac{5}{4}mg$ D $\frac{5}{3}mg$

26. The acceleration of P is

A $\frac{3}{5}g$ B $\frac{3}{4}g$ C $\frac{4}{5}g$ D $\frac{5}{4}g$

27. The speed of P is

A $\sqrt{\dfrac{g}{4a}}$ B $\sqrt{\dfrac{9ag}{4}}$ C $\sqrt{\dfrac{15ag}{4}}$ D $\sqrt{3ag}$

In question 28 a situation is described and is followed by several statements. Decide whether each of the statements is true (T) or false (F).

28. A particle P, of mass 0.5 kg, is travelling with constant angular speed $1.5\,\mathrm{rad\,s^{-1}}$ in a horizontal circle with centre O and radius 2 m.

   (i)   The velocity of P is constant.
   (ii) The speed of P is $3\,\mathrm{m\,s^{-1}}$.
   (iii) The acceleration of P is directed towards O.
   (iv) The acceleration of P is $4.5\,\mathrm{m\,s^{-2}}$.
   (v)   The time P takes to complete one orbit of the circle is $\frac{2}{3}\pi\,\mathrm{s}$.
   (vi) The resultant force on P is 2.25 N.

29. A light, inelastic string of length $2a$ is attached to fixed points A and B where A is vertically above B and the distance $AB < 2a$. A small, smooth ring P of mass m slides on the string and is moving in a horizontal circle at a constant angular speed ω. The string sections AP and PB are straight and there is the same tension T in each section. The distance AP is x and AP and PB make angles α and β respectively with the vertical, as shown in the diagram.

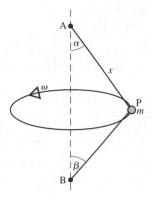

(i)   Show that $x \sin \alpha = (2a - x) \sin \beta$.
(ii) By considering the vertical components of forces on the ring, deduce why $x > a$.
(iii) By considering the radial motion of the ring, show that $T(\sin \alpha + \sin \beta) = mx\omega^2 \sin \alpha$.
(iv) Using your answer to (i), show that the tension in the string is $mx\omega^2(2a - x)/2a$. (MEI)

30.

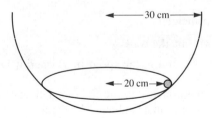

A small marble, of mass 5 grams, is moving in a horizontal circle of radius 20 cm on the smooth inner surface of a hemispherical bowl of radius 30 cm (see diagram). Find, in newtons, the magnitude of the force exerted on the marble by the bowl, and find also, in m s^{-1}, the speed of the marble. (UCLES)$_s$

31. A particle P, of mass M, moves on the smooth inner surface of a fixed hollow spherical bowl, centre O and inner radius r, describing a horizontal circle at constant speed. The centre C of this circle is at a depth $\frac{1}{3}r$ vertically below O. Determine

(a) the magnitude of the force exerted by the surface of the sphere on P,

(b) the speed of P. (ULEAC)

32. A car undergoing trials is moving on a horizontal surface around a circular bend of radius 50 m at a steady speed of $14\,\mathrm{m\,s^{-1}}$. Calculate the least value of the coefficient of friction between the tyres of the car and the surface. Find the angle to the horizontal at which this bend should be banked in order that the car can move in a horizontal circle of radius 50 m around it at $14\,\mathrm{m\,s^{-1}}$ without any tendency to side-slip.

Another section of the test area is circular and is banked at $30°$ to the horizontal. The coefficient of friction between the tyres of the car and the surface of this test area is 0.6. Calculate the greatest speed at which the car can move in a horizontal circle of radius 70 m around this banked test area.
(Take the acceleration due to gravity to be $10\,\mathrm{m\,s^{-2}}$.)   (AEB)

33. A steel ball B, of mass 0.125 kg, is attached to one end of a light elastic string OB, the end O being attached to the ceiling. The modulus of elasticity of the string is 52.5 N, and the string has natural length 1.5 m. In equilibrium the ball is at E. Show that the depth of E below O is 1.54 m, correct to 3 significant figures.

The ball is released from rest at O, and does not hit the floor. State an assumption necessary for conservation of energy to apply.

Hence find
  (i)   the speed as B passes through E,
  (ii) the maximum depth of B below O.   (UCLES)ₛ

34. A light elastic string of natural length a and modulus $7mg$ has a particle P of mass m attached to one end. The other end of the string is fixed to the base of a vertical wall. The particle P lies on a rough horizontal surface, and is released from rest at a distance $\frac{4}{3}a$ from the wall. The coefficient of friction between P and the surface is $\frac{1}{4}$.

Use the work-energy principle

(a) to show that P will hit the wall,

(b) to find, in terms of a and g, the speed of P when it hits the wall. (ULEAC)

35. In a charity event, a man is attached to the end of a light elastic rope, the other end of which is secured to a platform on a viaduct. The platform is 120 m above the ground. The natural length of the rope is 80 m and its modulus of elasticity is 17 640 N.

The man drops from the platform and falls without encountering any obstructions. (Air resistance may be neglected.) A 'safe' jump is one in which the man comes instantaneously to rest at least 10 m above the ground.

Using the principle of conservation of energy and treating the man as a particle,

(a) find, to the nearest kg, the mass of the heaviest man who can make a 'safe' jump,

(b) calculate the speed of a 75 kg man at a height of 20 m above the ground.
                                                              (ULEAC)

CHAPTER 4

MOTION IN A VERTICAL CIRCLE

MOTION ON A CURVE WITH VARIABLE SPEED

We saw in Chapter 2 that when a particle describes a curved path at constant speed, the particle has no acceleration in the direction of motion, i.e. no acceleration in the direction of the tangent to the curve at any instant. There is, however, an acceleration perpendicular to the direction of motion which is a measure of the rate of change of the direction of the velocity.

If we now consider motion on a curve when the speed is *not* constant it is clear that

(a)   again there is an acceleration component perpendicular to the tangential direction,

(b)   there is also an acceleration component *in* the direction of motion which is a measure of the rate of change of the magnitude of the velocity, and which can be expressed as $\mathrm{d}v/\mathrm{d}t$ (or $\dot{v}$).

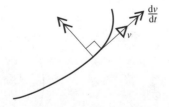

In order to produce these two acceleration components, the force acting on the particle must also have two components that are in the directions of the acceleration components.

This is the situation for a particle describing a curved path in a vertical plane; the weight always acts vertically downwards so, as the direction of motion changes, the weight can be resolved into two components, one along, and one perpendicular to, the direction of motion.

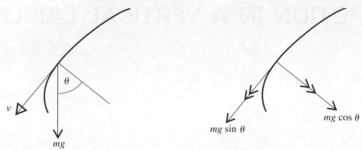

MOTION ON A CIRCLE IN A VERTICAL PLANE

When the path is a circle we know that the component of acceleration towards the centre is v^2/r or $r\omega^2$.

There are various ways by which an object can be made to travel in a vertical circular path. In some cases an object is controlled by a machine which involves technical knowledge beyond the scope of this book. In other cases the particle, once it is set moving, moves under the action of its own weight. It is situations of this type that we deal with here.

Motion Restricted to a Circular Path

Consider a small bead P, threaded on to a circular wire, radius a and centre O, that is fixed in a vertical plane. We will model the bead as a particle and the wire as being friction-free. Suppose that the bead has been set moving round the wire so that it passes through the lowest point A of the wire with speed u and that the speed of the bead is v when it reaches a point B on the wire, where angle AOB is θ.

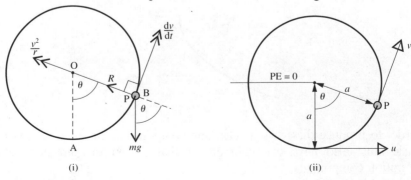

Diagram (i) shows the forces acting on the bead, and the acceleration components produced, as the bead passes through B.

Newtons Law, $F = ma$, can be applied towards O and along the tangent at B.

Using Newton's Law at B gives

$$\diagdown \quad R - mg \cos \theta = \frac{mv^2}{a} \qquad [1]$$

$$\diagup \quad mg \sin \theta = -m \frac{dv}{dt} \qquad [2]$$

The only external force that acts on the bead, other than its weight, is the normal reaction R. Now R is always perpendicular to the direction of motion so it does no work. Therefore conservation of mechanical energy can be applied and diagram (ii) is useful here.

Taking the PE to be zero at the level of the centre O, we have:

Total ME at A is $\quad \frac{1}{2} mu^2 - mga$

Total ME at B is $\quad \frac{1}{2} mv^2 - mga \cos \theta$

Using conservation of mechanical energy from A to B gives

$$\frac{1}{2} mu^2 - mga = \frac{1}{2} mv^2 - mga \cos \theta \qquad [3]$$

These equations provide a solution to most problems, in fact equations [1] and [3] are very often all that are required.

Note that the level of A could have been chosen as the PE zero level, but measuring heights relative to the level of the centre is often more straightforward.

Note also that the bead can move on the wire in one of two basic ways: it can perform complete circles or it can oscillate through an arc. In either case the path is at all times on a circle because the bead is physically prevented from leaving the wire.

Note also that the normal reaction acting on the bead can be either towards the centre of the circle or away from it, e.g.

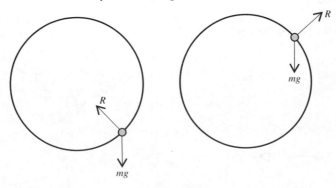

Another example of motion restricted to a circular path is that of a particle attached to one end of a rod that is rotating about the other end, which is pivoted at a fixed point. In this case it is the force in the rod which, together with the weight, causes the circular motion. This force, like the normal reaction in the case above, can act on the particle inwards (tension) or outwards (thrust).

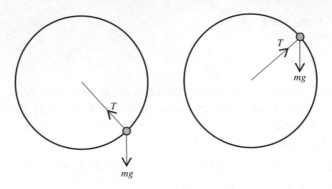

Example 4a

One end of a light rod of length a metres is pivoted at a fixed point O and a particle of mass m kg is attached to the other end. When the rod is hanging at rest, the particle is given a blow that makes it begin to move with velocity V metres per second.
Give answers in terms of a, m and g.

(a)   Find the value of V if the rod first comes to rest when horizontal.

(b)   (i)   Show that, for the particle to perform complete circles, $V \geqslant 2\sqrt{ga}$.
      (ii)   When $V = 2\sqrt{ga}$,   find the force in the rod when the particle is at the highest point, and say whether it is a tension or a thrust.

(c)   Given that $V = \sqrt{3ga}$,   find the height above O of the particle when the tension in the rod is zero.

The force acting on the particle towards the centre is the tension in the rod. It is always perpendicular to the direction of motion so does no work.

The diagrams show the particle at the lowest point A, and the velocities, forces and accelerations of the particle at a general point B where angle AOB is θ.

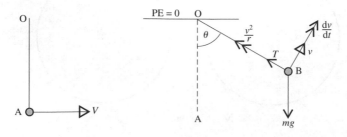

Using Newton's Law at B towards the centre gives

$$T - mg \cos \theta = \frac{mv^2}{a}$$ [1]

Using conservation of mechanical energy from A to B gives

$$\tfrac{1}{2}mV^2 - mga = \tfrac{1}{2}mv^2 - mga \cos \theta$$ [2]

(a) The rod first comes to rest when horizontal,

i.e. $v = 0$ when $\theta = 90°$

Hence from [2] $\quad \tfrac{1}{2}mV^2 - mga = 0 \quad \Rightarrow \quad V^2 = 2ga$

The value of V is $\sqrt{2ga}$.

(b) (i) For the particle to describe complete circles it must pass through the highest point with a positive velocity.

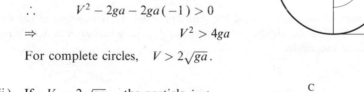

i.e. $v > 0$ when $\theta = 180°$

From [2] $\quad v^2 = V^2 - 2ga - 2ga \cos \theta$

$\therefore \quad V^2 - 2ga - 2ga(-1) > 0$

$\Rightarrow \quad V^2 > 4ga$

For complete circles, $V > 2\sqrt{ga}$.

(ii) If $V = 2\sqrt{ga}$, the particle *just* reaches the highest point C and the speed of the particle there is zero, i.e. v^2/a is zero.

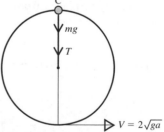

Applying Newton's Law at C gives

$$T + mg = \frac{mv^2}{a} = 0 \quad \Rightarrow \quad T = -mg$$

As T is negative the force in the rod acts outwards. Therefore the force in the rod is a thrust of magnitude mg.

(c) When $T = 0$,

from [1] $v^2 = -ga \cos \theta$

$\therefore$ from [2] $\frac{1}{2} v^2 - ga = -\frac{1}{2} ga \cos \theta - ga \cos \theta$

$\Rightarrow$ $V^2 - 2ga = -3ga \cos \theta$

As $V = \sqrt{3ga}$ $ga = -3ga \cos \theta$

$\Rightarrow$ $\cos \theta = -\frac{1}{3}$

The height of the particle above O is

$a \cos (180° - \theta)$

i.e. the particle is $\frac{1}{3} a$ above O.

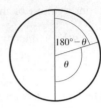

EXERCISE 4a

Questions 1 to 4 are about a bead P, of mass 1 kg, threaded on to a smooth circular wire, of radius 0.2 m, that is fixed in a vertical plane.

1. The bead is projected from the lowest point of the wire with a speed of $3\,\mathrm{m\,s^{-1}}$. Find the speed of P, $v\,\mathrm{m\,s^{-1}}$, when

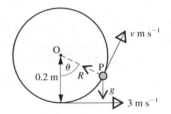

   (a) $\theta = 60°$ (b) $\theta = 90°$ (c) $\theta = 180°$

2. For each part of question 1, find the value of R where R newtons is the normal reaction of the wire on the bead. State in each case whether R is acting towards or away from the centre.

3. P is slightly displaced from rest at the highest point. Find v when

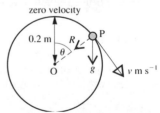

   (a) $\theta = 60°$ (b) $\theta = 90°$
   (c) $\theta = 180°$ (d) $\theta = 360°$

4. P is projected from the lowest point at $2.5\,\mathrm{m\,s^{-1}}$. Find the angle through which OP has rotated when the reaction between the wire and the bead is zero.

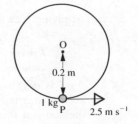

5. A particle P is fastened to one end of a light
rod of length 1.4 m. The other end of the rod is
smoothly pivoted to a fixed point O. The rod
is released from rest when OP is horizontal.
Find the speed of P, $v\,\mathrm{m\,s^{-1}}$, when OP has
rotated through   (a) 60°   (b) 90°.

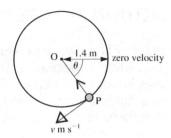

6. If in question 5 the mass of P is 1 kg, find for each angle the force T newtons
that the rod exerts on the particle, stating whether the force is a tension or a
thrust.

7. A smooth circular wire of radius 3 m is fixed in
a vertical plane. A bead P of mass 1.5 kg is
threaded on to the wire. P is projected at
5 m s⁻¹ from the highest point on the wire.
When the bead has rotated through 90°, find

   (a) the normal contact force exerted by the
        wire on the bead,

   (b) the force exerted on the bead in the
        direction of the tangent.

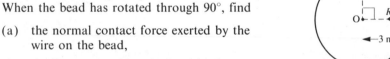

8. Referring to the situation described in question 7, when the bead has rotated
through 90°, find the radial and tangential components of the acceleration of P.

9. A light rod of length 3 m is smoothly pivoted at one end to a fixed point O. A
particle P of mass 1.5 kg is attached to the other end of the rod. The particle is
held so that OP is horizontal and is then projected downwards with speed 6 m s⁻¹.
Find the angle through which the rod has rotated when

   (a) the force exerted by the rod on the bead first becomes zero,

   (b) the speed of P first becomes zero.

   In each case find the height of P above O at that instant.

10. If in question 9 the rod is initially held at rest with P vertically above O and is
then slightly displaced, find the radial and tangential components of the
acceleration of P when OP has rotated through

   (a) 30°   (b) 90°   (c) 150°.

11. One end of a light rod of length a, is smoothly pivoted to a fixed point and a
particle of mass m is attached to the other end. When the rod is hanging
vertically downwards, the particle is set moving with a speed u. Find the values
of u for which the rod will rotate through a complete revolution.

MOTION NOT RESTRICTED TO A CIRCULAR PATH

So far we have considered situations where all the motion took place on a circular path, whether complete circles or oscillations were performed. There are other cases however in which an object begins by moving on a circle but then moves on a different path. Consider, for example, a particle fastened to one end of an inelastic string whose other end is fixed.

If the particle is set moving in a vertical plane it may, as in the cases considered earlier, rotate in complete circles

or it may oscillate through an arc that is less than a semicircle.

But there is now a third possibility. *If the string goes slack* during the motion, the particle will 'fall inside' the circular path and travel for a time under the action *only* of its weight.

The reason for the third case is that the string, unlike a light rod, cannot exert a thrust; it can only pull. So motion in a circle at the end of a string can take place only as long as the string is taut, i.e. as long as the tension, T, is greater than, or equal to, zero. At the instant when the particle is *about* to leave its circular path, $T = 0$.

If the particle is to describe complete circles we must ensure that $T \geqslant 0$ when $\theta = 180°$. (It is no longer sufficient to say that $v > 0$ when $\theta = 180°$, because this is true even if the particle has left the circular path.)

A particle set moving on the inside of a circular surface gives rise to a similar situation. The particle moves on a circle as long as it is in contact with the surface, i.e. as long as there is a positive inward contact force, R. If the particle loses contact with the surface its path falls inside the circle as shown above. At the point on the surface where contact is lost, $R = 0$.

If there is to be no loss of contact, i.e. complete circles are to be performed, then $R \geqslant 0$ at the top.

Examples 4b

1. A particle of mass m is attached to one end of a light inelastic string of length a whose other end is fixed to a point O. When the particle is hanging at rest it is given a horizontal blow which causes it to begin to move in a vertical plane with initial speed V. Find the ranges of values of V for which the particle at no time leaves a circular path.

The particle will move in a circular path provided that the string does not go slack.

One situation in which the string cannot go slack is when the particle oscillates through no more than 180°

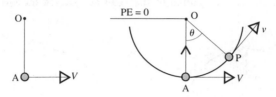

The tension is always perpendicular to the direction of motion so it does no work.

Using conservation of mechanical energy from A to P,

$$\tfrac{1}{2}mV^2 - mga = \tfrac{1}{2}mv^2 - mga\cos\theta \qquad\qquad [1]$$

For oscillations through not more than 180°,

$$v = 0 \quad \text{when} \quad \theta \leqslant 90°, \quad \text{i.e. when} \quad \cos\theta \geqslant 0$$

Hence, from [1] $\qquad mga\cos\theta = mga - \tfrac{1}{2}mV^2 \geqslant 0$

$$\Rightarrow \qquad V \leqslant \sqrt{2ga}$$

The other situation in which the string does not go slack is when the particle describes complete circles. For this to happen the tension must be greater than or equal to zero when $\theta = 180°$.

Using $\theta = 180°$ in [1] gives $\tfrac{1}{2}mV^2 - mga = \tfrac{1}{2}mv^2 - mga(-1)$

$$\Rightarrow \qquad\qquad v^2 = V^2 - 4ga$$

Using Newton's Law towards the centre at the highest point gives

$$T + mg = \frac{mv^2}{a} \qquad \Rightarrow \qquad T = \frac{m}{a}(V^2 - 4ga) - mg$$

When $\theta = 180°$, $T \geqslant 0 \quad \Rightarrow \quad \dfrac{m}{a}(V^2 - 4ga) - mg \geqslant 0$

$$\Rightarrow \qquad V \geqslant \sqrt{5ga}$$

$\therefore$ The particle will not leave the circular path if $V \leqslant \sqrt{2ga}$ or $V \geqslant \sqrt{5ga}$.

2. A smooth cylinder of radius 0.3 m is fixed, with its axis horizontal, on a horizontal plane. A small bead of mass 1 kg is placed at the highest point A on the outside of the cylinder and is just displaced from rest. The bead subsequently loses contact with the surface of the cylinder at a point P.

(a)   Find the height of P above the axis of the cylinder.

(b)   State, to the nearest degree, the angle between the horizontal and the direction of motion of the bead when it leaves the cylinder.

(c)   Find the speed and the tangential acceleration of the bead at this instant.

(d)   Describe the subsequent motion of the bead and sketch its path.

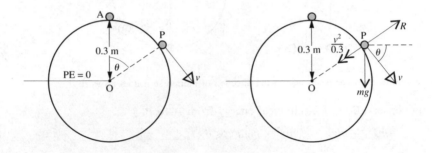

(a)   The normal reaction R acts outward.

The bead will leave the surface when $R = 0$.

Conservation of ME from A to P gives

$$0 + (1)(9.8)(0.3) = \tfrac{1}{2}(1)v^2 + (1)(9.8)(0.3\cos\theta)$$

$$\Rightarrow \qquad\qquad v^2 = 5.88(1 - \cos\theta) \qquad\qquad\qquad [1]$$

Using Newton's Law at P gives

$$(1)(9.8)\cos\theta - R = (1)\left(\frac{v^2}{0.3}\right) \qquad\qquad\qquad [2]$$

$$= 19.6(1 - \cos\theta) \qquad\qquad \text{using } [1]$$

When $R = 0$, $29.4\cos\theta = 19.6$ $\Rightarrow$ $\cos\theta = \tfrac{2}{3}$

The height of the bead above the centre is $0.3\cos\theta$

The bead leaves the surface of the cylinder when it is at the point which is 0.2 m above the centre.

(b)   Using Newton's Law along the
      tangent gives

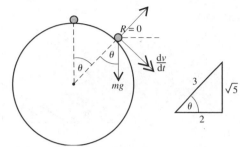

$$mg \sin \theta = m \frac{dv}{dt}$$

$$\Rightarrow \qquad \frac{dv}{dt} = \frac{g\sqrt{5}}{3} = 7.304\ldots$$

The tangential acceleration is
$7.30 \, \text{m s}^{-2}$ (3 sf).

(c)   From (a) $\cos \theta = \frac{2}{3}$ $\Rightarrow$ $\theta = 48°$ (to the nearest degree).

Therefore the direction of motion of the bead when it leaves the surface of
the cylinder is at $48°$ to the horizontal.

From [1], the speed of the bead when it leaves the surface is given by

$$v^2 = 5.88 \left(1 - \tfrac{2}{3} \right) \qquad \Rightarrow \qquad v = 1.4$$

The speed of the bead is $1.4 \, \text{m s}^{-1}$.

(d)   The bead now moves under the action of
      its weight only, therefore it travels as a
      projectile with initial speed $1.4 \, \text{m s}^{-2}$, at
      $48°$ to the horizontal.

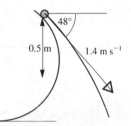

3. **The diagram shows a section of a toy racing car track. It consists of a slope of
    length l metres, inclined at $30°$ to the horizontal, which levels off and then curves
    upward and round in a complete circle of radius 25 cm, in which the car 'loops the
    loop'.**

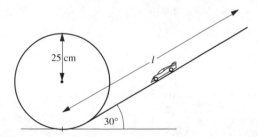

(a) **Making an assumption, which you should state, use energy considerations to
     find, in terms of l and g, the velocity V of the car at the foot of the incline,
     given that it starts from rest at the top.**

(b) **The car of mass m kg is intended to travel round the inside of the circular loop
     without losing contact with the track. By defining an appropriate model, show
     that the value of l must be at least 1.25.**

(a)

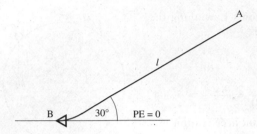

Assuming that there is no resistance to the motion of the car, we can use conservation of mechanical energy from A to B.

$$mgl \sin 30° + 0 = 0 + \tfrac{1}{2}mV^2$$

$$\Rightarrow \qquad V = \sqrt{gl}$$

(b) Model the car as a particle and the circular track as smooth. Assume that the velocity of the car at lowest point on the circular loop is equal to that at the foot of the incline.

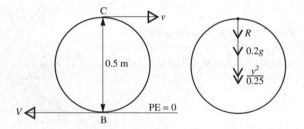

The normal reaction R does no work.

Using conservation of ME from B to C gives

$$\tfrac{1}{2}mV^2 + 0 = \tfrac{1}{2}mv^2 + m(9.8)(0.5)$$

$$\Rightarrow \qquad \tfrac{1}{2}gl = \tfrac{1}{2}v^2 + 4.9$$

$$\Rightarrow \qquad v^2 = 9.8(l-1) \qquad\qquad\qquad [1]$$

If the car loses contact with the track, it will be when the contact force disappears, so the condition we want is that $R \geqslant 0$ at the highest point.

Using Newton's Law towards the centre at C gives

$$R + m(9.8) = m\left(\frac{v^2}{0.25}\right)$$

$$\Rightarrow \qquad R = 0.8v^2 - 1.96$$

$$R \geqslant 0 \qquad \Rightarrow \qquad v^2 \geqslant 2.45$$

$$\therefore \quad \text{from [1]} \qquad 9.8(l-1) \geqslant 2.45$$

$$\Rightarrow \qquad\qquad l \geqslant 1.25$$

The least value of l is 1.25.

EXERCISE 4b

Questions 1 to 3 are about a particle P fastened to one end of a light inextensible string whose other end is fixed at a point O.

1. When P, whose mass is 0.2 kg, is hanging vertically below O, it is given a horizontal speed of 5 m s^{-1}. If the length of the string is 0.5 m, find the speed of P when OP has rotated through 60°.

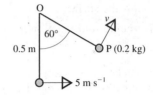

2. If, in question 1, OP is initially horizontal and P is then given a speed 3 m s^{-1} vertically downwards find

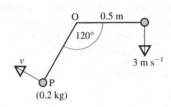

   (a) P's speed when OP has rotated through 120°,

   (b) the tension in the string at that instant.

3. From a position where the string, of length 1 m, is horizontal and taut, P is given an initial speed u m s^{-1} vertically downwards.

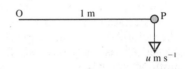

   (a) Find the range of values of u for which P will describe complete circles.

   (b) Describe the motion of P if $u = 0$.

In questions 4 to 6 a particle P is set moving in a vertical plane on the inside smooth surface of a cylinder.

4. P is at the lowest point on the surface when it is given a horizontal speed of 6 m s^{-1}. The mass of the particle is 0.5 kg and the radius of the cylinder is 1 m. After P has rotated through 70°, find its speed and the normal reaction exerted by the surface on P.

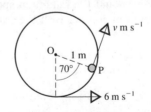

5. P is projected horizontally from the lowest point of the surface, whose radius is a, and after rotating through 90° its speed is $\sqrt{7ga}$. Find the initial speed in terms of a and g.

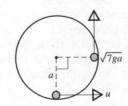

6. The particle, of mass m, is projected vertically upwards from a point level with the centre of the cylinder, with speed $\sqrt{7ga}$. Find

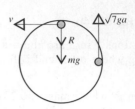
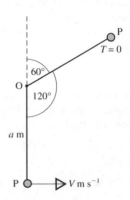

   (a) the speed when P reaches the highest point on the surface,

   (b) the normal reaction at that point.

7. A particle P is hanging at rest at the end of a light inextensible string of length a m. The other end of the string is fixed at a point O. P is then projected horizontally with speed V m s^{-1}. When OP has rotated through 120°, the string becomes slack (i.e. the tension is zero). Giving answers in terms of a and g find,

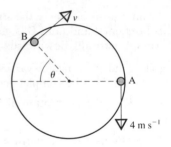

   (a) the speed of P at this instant

   (b) the value of V.

8. A particle is held in contact with the smooth inner surface of a cylinder of radius 0.08 m, at the point A as shown. It is then given a speed of 4 m s^{-1} vertically downwards and rotates until, at point B, the normal reaction becomes zero. Find the height of B above A and the speed of the particle at B.

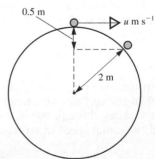

9. An aircraft is looping the loop on a path which is a vertical circle of radius 400 m. Find the minimum speed at the top of the loop for which the pilot would remain in contact with the seat without wearing a seat belt.

   Questions 10 and 11 are about a particle P, set moving in a vertical plane on the smooth outer surface of a fixed sphere.

10. The mass of P is 1 kg and the radius of the sphere is 2 m. P is projected horizontally with speed u m s^{-1} from the highest point on the sphere and loses contact with the surface when it has descended a vertical distance 0.5 m. Find the value of u.

11. The sphere, whose radius is 2 m, is fixed on a horizontal plane and P is just displaced from rest at the highest point. Find

 (a) the height of P above the plane when contact with the sphere is lost,

 (b) P's speed at this instant,

 (c) P's speed when it reaches the plane (use an energy method).

12. A child whirls a basket containing an apple, of mass m kilograms, in a vertical circle of radius 0.6 m. The speed of the apple at the highest point is v metres per second.

 (a) Find, in terms of m and v, the force exerted by the basket on the apple when the apple is at the highest point.

 (b) Find the least value v can have if the apple is not to lose contact with the basket.

*13. One very cold winter's day the dome of St Paul's Cathedral becomes icy. A pigeon tries to land gently on the dome but immediately it comes to rest on the surface it finds itself sliding down. Surprised, but interested in this new experience, it sits tight until it loses contact and then flies away. The dome is to be modelled as a hemisphere, of radius 15 m, and its icy surface as frictionless. The pigeon's initial position on the roof is at an angular displacement of 10° from the highest point of the hemisphere.

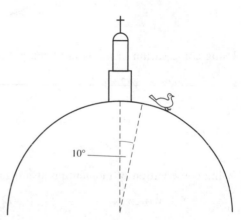

 (a) Find the distance it slides over the roof before losing contact.

 (b) Find its speed just before it starts to fly.

HARDER PROBLEMS

There are many different types of situation involving motion on a circular path in a vertical plane, some requiring ideas that have not been used so far in this chapter. The examples that follow give an indication of the variety of questions that you might meet and the next exercise includes more of them.

In some questions you will need to use simple applications of impulse and momentum. If you have forgotten the principles involved you will find them in a revision section at the beginning of Chapter 5 and should work through that section before attempting these problems.

Examples 4c

1. A magnet of mass $2m$ kg is attached to one end of a string of length a m. The other
    end of the string is fixed at a point A. The magnet is held, with the string taut, at
    a point B level with A, and is released from rest from that position. When the
    magnet is at the lowest point of its motion it picks up a stationary iron block of
    mass m kg. Find the height to which the combined mass rises.

First find the speed of the magnet just before it picks up the iron block.

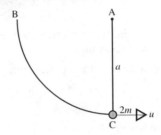

Using conservation of mechanical energy from B to C gives

$$2mga = \tfrac{1}{2}(2m)u^2 \qquad \Rightarrow \qquad u = \sqrt{(2ga)}$$

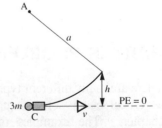

Using conservation of momentum at impact gives

$$2mu + 0 = 3mv \qquad \Rightarrow \qquad v = \tfrac{2}{3}u = \tfrac{2}{3}\sqrt{2ga}$$

Now we will consider energy again as the mass $3m$ rises to an unknown height h m above C.

Using conservation of mechanical energy
from C to D gives

$$\tfrac{1}{2}(3m)v^2 = 3mgh$$

$$\Rightarrow \qquad h = \frac{v^2}{2g} = \left(\frac{1}{2g}\right)\left(\frac{8ga}{9}\right)$$

The combined mass rises to a height of $\tfrac{4}{9}a$.

In most of the problems in this chapter an object has been travelling on a vertical
circle under the action of its own weight, and therefore with varying speed. There
are situations however in which a vertical circular path is described at constant
speed, e.g. the motion of a 'Big Wheel' once all the passengers are aboard.
The next example illustrates the special features of this case.

2. A thin metal spoke, of length 1.2 m, carries a small
load of mass 0.7 kg at one end and is pivoted at the
other end to a fixed point O. The spoke is being
rotated manually in a vertical plane at a constant
angular speed of 1 revolution per second.

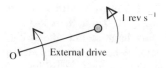

(a) Explain why the conservation of mechanical energy cannot be used.

(b) What is the acceleration of the load towards O when the spoke is
(i) horizontal   (ii) vertically above O   (iii) vertically below O.

(c) Find the force exerted by the spoke on the load in each of the positions defined
in part (b), stating whether the spoke is in tension or compression.

(a) External work is being done by whatever is producing the manual rotation so
as to maintain a constant angular speed. This causes a change in mechanical
energy, so the total mechanical energy of the system is not constant.
(Note that the external drive must be a torque (turning effect) and not a
linear force)

(b) The angular speed of the spoke and load is $1 \, \text{rev s}^{-1}$, i.e. $2\pi \, \text{rad s}^{-1}$. The
angular speed of the load, and the radius of the circle, are the same at every
point so $r\omega^2$, the central acceleration, is also the same at every point.
Therefore in each of the given positions

$$\text{the acceleration towards O is} \quad 1.2 \times (2\pi)^2 \, \text{m s}^{-2}$$

$$\text{i.e.} \quad 47 \, \text{m s}^{-2} \quad (2 \, \text{sf})$$

(c)

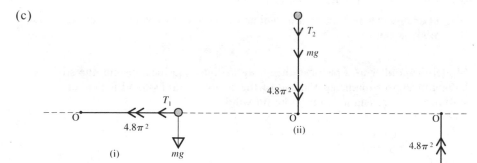

We will apply Newton's Law towards O in each case.

(i) $T_1 = (0.7)(4.8\pi^2) = 3.36\pi^2 = 33.16\ldots$

∴   the tension is 33 N (2 sf).

(ii) $T_2 + 0.7g = 3.36\pi^2$ ⇒ $T_2 = 26.3\ldots$

∴   the tension is 26 N (2 sf).

(iii) $T_3 - 0.7g = 3.36\pi^2$ ⇒ $T_3 = 40.02\ldots$

∴   the tension is 40 N (2 sf).

3. Paul is playing with a conker of mass 30 g, fastened to the end of a string 20 cm long. He holds the end of the string in one hand A and holds the conker, C, in the other hand so that the string is taut with AC horizontal. When Paul releases the conker, he intends to watch and see whether it goes right round to be level with his stationary hand again but, after he has let the conker go, one of his friends pushes a stick in the way so that, just as the string becomes vertical, the middle of the string hits the stick at right angles. The conker continues to rotate about the stick as centre.

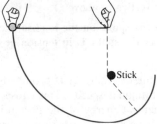

Make a mathematical model for this situation, stating the assumptions made, and use it to

(a)   find the speed of the conker just before the string hits the stick and the tension in the string at this instant

(b)   explain why the speed of the conker does not change when the string hits the stick

(c)   find the tension in the string immediately after it hits the stick

(d)   determine whether or not the conker will describe a complete circle about the stick as centre.

Model the conker as a particle, the string as light and inextensible, the stick as having negligible diameter, the end of the string in Paul's hand is perfectly stationary, no resistance to motion, no wind.

(a)

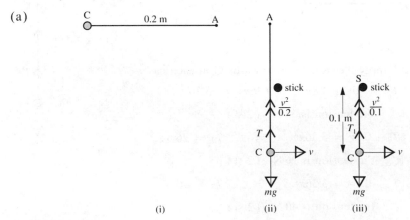

Using conservation of mechanical energy from (i) to (ii)

$$0.03 \times 9.8 \times 0.2 = \left(\tfrac{1}{2}\right)(0.03)\,v^2 \qquad \Rightarrow \qquad v^2 = 3.92$$

$$\Rightarrow \qquad v = 1.98 \quad (3 \text{ sf})$$

Using $F = ma$ in (ii) gives

$$T - mg = \frac{mv^2}{r}$$

$$\Rightarrow \qquad T = (0.03)\left(\frac{3.92}{0.2} + 9.8\right)$$

The tension is 0.88 N (2 sf)

(b) When the string strikes the stick, the radius of the circle being described changes suddenly causing a change in the central acceleration and hence in the tension. However, this instantaneous change in tension is perpendicular to the direction of motion of the conker so the speed of the conker is unchanged.

(c) The radius of the circle being described is now 0.1 m so the tension changes.

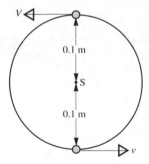

$$T_1 - mg = \frac{mv^2}{r}$$

$$\Rightarrow \qquad T_1 = (0.03)\left(9.8 + \frac{3.92}{0.1}\right)$$

The new tension is 1.5 N (2 sf).

S

0.1 m

T_1

C

v

mg

(d) The conker will describe a complete circle about the stick provided that the conker *passes through* the highest point of that circle, i.e. after 180° rotation the speed is not zero, and the tension is positive or zero.

V

0.1 m

S

0.1 m

v

Conservation of ME gives $\qquad \tfrac{1}{2}mv^2 - mg(0.1) = \tfrac{1}{2}mV^2 + mg(0.1)$

$$\Rightarrow \qquad V^2 = v^2 - 0.4g = 3.92 - 3.92 = 0$$

∴   the conker comes to rest vertically above the stick and will drop straight down. So, without investigating the tension, we know that the conker will not complete a circle.

(Note that the conker cannot already have 'fallen inside' the circle as in that situation the speed cannot be zero.)

EXERCISE 4c

1. One end of a light rod of length 3 m, is smoothly pivoted
   to a fixed point O and a particle P of mass 1.5 kg is
   attached to the other end. When the rod is hanging at
   rest, vertically downwards, P is given a horizontal blow of
   impulse 18 N s as shown in the diagram.

   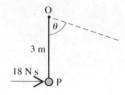

   (a) Find its initial velocity.

   (b) Show that P moves round complete circles.

   (c) Find the force in the rod, T N, when

       (i) $\theta = 160°$ (ii) $\theta = 170°$,

       stating in each case whether it is a tension or a thrust.

   (d) Find the value of θ when $T = 0$.

2. One end of a light rod of length 0.4 m is pivoted at a fixed point A so that it can
   swing freely in a vertical plane. A particle B of mass 1 kg is attached to the other
   end of the rod. The rod is released from rest with AB horizontal. When B is
   vertically below A it collides with a particle C, of mass 0.5 kg, which is
   approaching it horizontally at $2 \, \text{m s}^{-1}$ and the two particles coalesce. Find

   (a) the velocity of B and the tension in the rod just before impact

   (b) the common velocity just after impact

   (c) the tension in the rod just after impact

   (d) the height above the lowest point to which the combined particles now
   swing.

3. A car, of mass 800 kg, is driven
   over a hump-backed bridge at a
   constant speed of $14 \, \text{m s}^{-1}$. The
   upper part of the bridge can be
   modelled as an arc of a circle, of
   radius 25 m, this arc subtending an
   angle of 30° at the centre of the
   circle.

   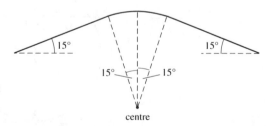

   The gradient of the road approaching the arc is equal to tan 15°. Find

   (a) the reaction on the car when it is at the highest point of the bridge

   (b) the reaction on the car when it is at the end of the curved section of the
   bridge

   (c) the reaction on the car when it is on the straight section

   (d) the greatest constant speed at which a car may be driven over this bridge
   without losing contact with the road.

4. A particle of mass 1.5 kg is lying at the lowest point of the einner surface of a hollow sphere of radius 0.5 m when it is given a horizontal impulse. Find the magnitude of the impulse:

    (a)   if the particle subsequently describes complete vertical circles,

    (b)   if the particle loses contact with the sphere after rotating through 120°.

5. One end of a light inextensible string AB of length l is fixed at A and a particle of mass m is attached at B. B is held a distance l vertically above A and is projected horizontally from this position with speed $\sqrt{2gl}$. When AB is horizontal, a point C on the string strikes a fixed smooth peg so that the radial acceleration of the particle is instantaneously doubled. Express the length of CB in terms of l.

    The particle continues to describe vertical circles about C as centre. Compare the greatest and least tensions in the string during this motion.

6. A pair of trapeze artists are performing their act. There is a catcher and a flier. The catcher, of mass 75 kg swings on an arc of radius 7 m and his speed is zero when the trapeze ropes are at 70° to the vertical. When he reaches the lowest point of his path he catches the flier, who has a mass of 55 kg and is approaching him with a horizontal velocity of $2\,\mathrm{m\,s^{-1}}$. Stating all assumptions which you make, find

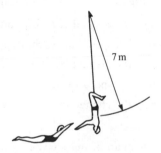

    (a)   the speed of the catcher just before they connect,

    (b)   their common velocity just after connecting,

    (c)   the force with which the catcher must grip the trapeze just after connecting,

    (d)   the angle the trapeze ropes will make with the vertical when they first come to instantaneous rest.

7. A ballistic pendulum is an instrument used to measure the speed of a bullet. In this problem such a pendulum consists of a bob, of mass 5 kg, suspended by a thin light rod from a fixed point. The bob swings on an arc of a circle of radius 0.6 m against a scale which measures its rotation from the vertical. While the pendulum is hanging vertically at rest a bullet, of mass 0.01 kg is fired horizontally into it. The bob with the bullet embedded in it is observed to swing to a maximum deflection of 24° from the vertical.

    (a)   Find the common velocity of the bob and bullet just after impact.

    (b)   Find the tension in the rod just after impact.

    (c)   Find the velocity of the bullet before hitting the bob.

8.

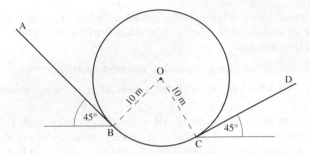

A boy is using his sledge on a snow-covered hillside, which is to be modelled as frictionless. The total mass of the boy and sledge is 50 kg. He starts from rest at point A and slides down the slope AB, which is inclined at 45° to the horizontal and is 20 m long. He then goes into a hollow BC, which is to be treated as an arc of a circle, of radius 10 m, to which AB is a tangent. He comes out of this hollow, at point C, on to slope CD, which is also a tangent to the arc BC and is inclined at 30° to the horizontal. Find

(a)   the normal reaction and his acceleration while on section AB

(b)   his speed on reaching point B

(c)   the normal reaction just after entering arc BC

(d)   his speed on reaching point C

(e)   the change in normal reaction as he passes point C.

9. A woman of mass 60 kg is riding on the Big Wheel at a fairground. She is moving on a circle of radius 8 m at a constant angular speed of 5 revolutions per minute. When she has an angular displacement θ from the lowest point, the seat exerts a force on her with components R towards the centre and S along the tangent. Find, in terms of π and θ

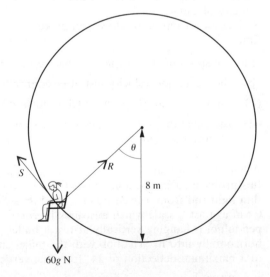

(a)   her speed in $\mathrm{m\,s^{-1}}$

(b)   the components of her acceleration
    (i)   towards the centre
    (ii)   along the tangent

(c) R and S

(d)   Evaluate, correct to two significant figures, the magnitude of the resultant of R and S when   (i) $\theta = 0°$ (ii) $\theta = 180°$,   and state the direction of this resultant.

10. A light aircraft is looping the loop on a path which is a vertical circle of radius 300 m. At a particular point on its path its speed is $60 \, \mathrm{m \, s^{-1}}$ and this speed is increasing at the rate $9 \, \mathrm{m \, s^{-2}}$. Find

(a)   its acceleration towards the centre of the circle

(b)   the magnitude of its resultant acceleration.

11. A building is being demolished by swinging a large metal ball at the walls. The ball has a mass of 500 kg and it is attached by a chain to a fixed point. It swings on an arc of a circle of radius 6 m. A cable draws the ball back to a position where the chain is inclined at 30° to the vertical and the ball is then released. It strikes a wall when the chain reaches the vertical position and the impact brings it to rest.

(a)   Find the speed of the ball when it strikes the wall.

(b)   Find the impulse exerted on the wall.

(c)   Find the tension in the chain
   (i)   just before the ball strikes the wall
   (ii) just after the ball strikes the wall.

12. A particle lies at the lowest point on the inside surface of a smooth cylinder, of radius 0.9 m, which is fixed with its axis horizontal. It is then given a horizontal velocity so that it starts to move in a vertical circle. It leaves the surface on reaching point A, after rotating through 120° from the lowest point.

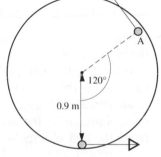

(a)   Find its velocity on reaching point A.

(b)   Find its initial velocity.

(c)   Find the horizontal and vertical components of its velocity at the moment it loss contact.

(d)   By using the vertical component, find the greatest height reached above point A.

(e)   Sketch the path traced out by the particle after leaving the surface.

*13. A particle lies on the outer surface of a smooth sphere, of radius 1 m, which is standing on horizontal ground. It is projected from the highest point with a velocity of $2.5 \, \mathrm{m \, s^{-1}}$.

(a)   Find the angle the particle has rotated through when it loses contact with the surface.

(b)   Find its height above the ground at the moment it loses contact.

(c)   Find its velocity at the moment it loses contact.

(d)   Find the horizontal and vertical components of its velocity at the moment it loses contact.

(e)   Consider its motion as a projectile after leaving the surface and find
   (i)   the vertical component of its velocity on reaching the ground
   (ii) the time taken to reach the ground
   (iii) the horizontal distance it has travelled in this time.

CHAPTER 5

COLLISIONS
LAW OF RESTITUTION

BASIC IDEAS OF MOMENTUM

In Module E the following definitions and properties were established.

● The momentum of an object of mass m, moving with velocity v, is of magnitude mv. Because mass is scalar and velocity is vector, mass × velocity is also vector, i.e. momentum is a vector quantity.

● When a constant force F acts for a time t on an object of mass m, a change in the momentum of the object is produced, of magnitude Ft. The product Ft is a vector quantity and is called the impulse of the force. The unit in which it is measured is the newton second, N s.
If an impulse Ft causes the velocity of the object to increase from u to v, then   Impulse = Increase in momentum,   i.e. $Ft = mv - mu$,
hence the unit of impulse is also used as the unit for momentum.

● If an unknown force acts for an unknown time, so that $F \times t$ cannot be calculated, the impulse of the force can be found from the change in momentum it produces.   This situation exists when two moving objects collide and the impulse in such cases is called an instantaneous impulse.

● When two objects collide, each exerts an instantaneous impulse on the other. These impulses are equal and opposite so, for the pair of objects, there is no resultant impulse and hence no overall change in momentum. This property is known as the principle of conservation of linear momentum and is expressed formally as follows.

If no external impulse acts on a system of moving objects,
the total momentum in any specified direction remains constant.

We recommend that this work be revised before continuing; the following exercise can be used for this purpose.

EXERCISE 5a

In questions 1 to 3 a particle of mass 2 kg, moving along a straight line AB at $5 \, \text{m s}^{-1}$, receives an impulse.

1. After the impulse acts, the velocity is $8 \, \text{m s}^{-1}$ in the direction AB. Find the magnitude and direction of the impulse.

2. If the impulse is 6 N s in the direction BA, find the magnitude and direction of the velocity after the impulse.

3. If the impulse is 12 N s in the direction BA, find the magnitude and direction of the velocity after the impulse.

4. A particle of mass 0.5 kg moving with velocity $2\mathbf{i} + 5\mathbf{j} \, \text{m s}^{-1}$ receives an impulse $5\mathbf{i} - 7\mathbf{j} \, \text{N s}$. After the impulse acts find

   (a) the velocity of the particle,   (b) its speed.

5. A ball approaches a tennis player at $30 \, \text{m s}^{-1}$ and she returns it at $40 \, \text{m s}^{-1}$ along the same line. The ball has a mass of 55 g and is in contact with the racket for 0.04 s. Calculate the average force exerted on the ball during this time.

6. A ball of mass 0.2 kg strikes a wall at right angles with a speed of $8 \, \text{m s}^{-1}$. After rebounding it has 25% of its initial kinetic energy.

   (a) Find its speed after rebounding.

   (b) Find the impulse it exerts on the wall.

   (c) If it is in contact with the wall for 0.01 s, find the average force it exerts on the wall.

7. A spacecraft, with the final stage of its rocket still attached, is travelling at $16\,000 \, \text{m s}^{-1}$ when the rocket stage is detached by exploding a charge between it and the spacecraft. The mass of the spacecraft is $4.5 \times 10^4 \, \text{kg}$ and the mass of the rocket stage is $8 \times 10^4 \, \text{kg}$. The explosion gives an impulse of $2.25 \times 10^6 \, \text{N s}$ to each part. Find the magnitudes and directions of the velocities of the two parts after separation.

8. A railway truck of mass 36 tonnes travelling at $1.5 \, \text{m s}^{-1}$ collides with another truck of mass 24 tonnes travelling at $1 \, \text{m s}^{-1}$. They move coupled together after impact. Find the speed after impact and the loss in kinetic energy

   (a) if they are moving in the same direction before colliding

   (b) if they are moving towards each other before colliding.

9. An empty punt of mass 50 kg is drifting down a river at $2\,\mathrm{m\,s^{-1}}$. It is approached by a second punt moving at $3\,\mathrm{m\,s^{-1}}$ and, when they are level, a man of mass 70 kg steps across sideways into the empty punt. Find the velocity that he and his new punt have just after he does this

   (a) if he approaches by overtaking the empty punt

   (b) if he approaches from the opposite direction.

10. A gas molecule having a velocity $300\mathbf{i}\,\mathrm{m\,s^{-1}}$ collides with another molecule of the same mass which is initially at rest. After the collision the first molecule has velocity $225\mathbf{i} + 130\mathbf{j}\,\mathrm{m\,s^{-1}}$. Find, in the form $a\mathbf{i} + b\mathbf{j}$, the velocity of the second molecule after the collision.

11. Two skaters, father and son, are standing at rest on the ice. The father then slides a stone over the ice giving it a velocity of $8\,\mathrm{m\,s^{-1}}$. The son catches it. The masses are: father 70 kg, son 50 kg and stone 5 kg.

   (a) Find the speed with which the father starts to move backwards after releasing the stone.

   (b) Find the common velocity of the son and the stone after he has caught it.

   (c) Find the impulse exerted by the stone on the son.

   State any assumptions you need to make to solve this problem.

ELASTIC IMPACT

Most of the collisions that have been considered so far have resulted in the objects coalescing at impact. Impacts of this type are called *inelastic*.

If, on the other hand, a bounce occurs at collision, we have an *elastic impact* and the colliding object(s) are said to be *elastic*.

Consider a particle that is moving on a horizontal plane.

The simplest example of an elastic impact is a *direct* impact, i.e. an impact in which the direction of motion just before impact is parallel to the impulses that act at the instant of collision, e.g.

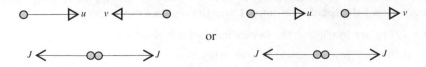

or

NEWTON'S LAW OF RESTITUTION

When two particles approach each other, collide directly and then move apart, the speed with which they separate is usually less than the speed at which they approached each other.

Experimental evidence suggests that, for two particular colliding particles, the separation speed is always the same fraction of the approach speed. It was from such evidence that Newton formulated a law known as Newton's *Law of Restitution*, which states that

$$\text{separation speed} = e \times \text{approach speed}$$

or

$$\text{relative speed after impact} = e \times \text{relative speed before impact}$$

The quantity represented by e is called the *coefficient of restitution* and it is constant for any two particular objects; its value depends upon the materials of which the two objects are made.

For colliding particles, e can take any value from zero to 1.

If the particles coalesce the separation speed is zero, i.e. $e = 0$.

If the relative speeds before and after impact are equal, $e = 1$, and we say that the particles are *perfectly elastic*; we shall see later on that in this case there is no loss in kinetic energy because of the impact.

Note that we have been referring only to particles colliding. A particle is regarded as having no measurable size, which means that its shape cannot be distorted and the time for which each particle is in contact with the other is infinitesimal. A larger object however, such as a ball, can undergo distortion on impact and the internal changes in its structure may be such that the basic methods for collision are not completely accurate. It is, however, reasonable to model most balls as elastic particles, as the results are accurate enough for most purposes.

Note also that when we refer to the speeds before and after impact, we mean the speeds *immediately* before and after. If, for instance, a particle falls vertically from a height above a fixed plane, its speed increases as it falls; the speed with which it *approaches the plane* is understood to be the final speed a split second before the collision with the plane.

Similarly the speed of separation from the plane is the initial speed with which the particle begins to rise again.

COLLISION WITH A FIXED OBJECT

Consider first the case of a particle of mass m, moving on a smooth horizontal surface with speed u, towards a fixed block whose face is perpendicular to the direction of motion of the particle. When the particle hits the block an impulse J is exerted on the particle by the block and, if the impact is elastic, the particle bounces off the block in the opposite direction with speed v, say.

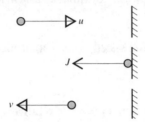

The approach speed is u, and the separation speed is v.

Therefore using the law of restitution gives

$$v = eu$$

Now, taking the direction of J as positive,

using   impulse = final momentum – initial momentum   gives

$$J = mv - (-mu)$$

These are the two principles that can be applied to situations of this type in which one of the colliding objects is fixed. The conservation of linear momentum is not valid in such cases for the impulse applied to the particle by the fixed surface is an *external* impulse; hence the momentum of the particle is changed but the momentum of the fixed object is not changed by an equal and opposite amount.

Examples 5b

1. **During a game of squash the ball strikes the front wall at right angles, at a speed of $20\,\text{m}\,\text{s}^{-1}$. At the moment of impact the ball is travelling horizontally. The coefficient of restitution between the ball and the wall is 0.9. By modelling the squash ball as an elastic particle, find**

   (a)   the speed with which the ball bounces off the wall

   (b)   the impulse exerted on the ball by the wall, given that the mass of the ball is $0.085\,\text{kg}$.

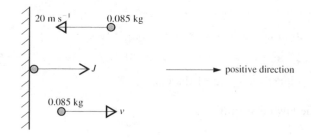

(a)   Using the law of restitution, $v = 0.9 \times 20 = 18$

The ball leaves the wall at $18\,\mathrm{m\,s^{-1}}$.

(b)   Using impulse $=$ increase in momentum in the direction of J,

$$J = mv - m(-20) = 0.085(18 + 20)$$

The impulse exerted is $3.23\,\mathrm{N\,s}$.

2. **A ball of mass 0.15 kg is dropped from a height of 2.5 m above horizontal ground. After bouncing it rises to a height of 1.6 m.**
    Stating any assumptions you need to make, find

    (a)   the coefficient of restitution between the ball and the ground

    (b)   the ball's loss in mechanical energy caused by the impact with the ground and suggest a possible explanation for this loss.

Assumptions are: the ball can be regarded as an elastic particle; there is no air resistance or wind; the ball rises vertically after impact with the ground.

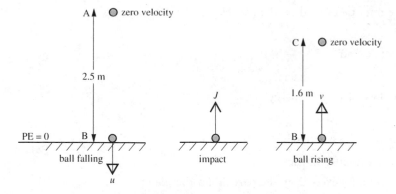

First find the speed of the ball at impact.

Using conservation of mechanical energy from A to B

$$mgh + 0 = 0 + \tfrac{1}{2} mu^2 \qquad \Rightarrow \qquad u^2 = 2 \times 9.8 \times 2.5$$

$$\Rightarrow \qquad u = 7$$

$\therefore$ the speed of the ball just before impact is $7\,\mathrm{m\,s^{-1}}$.

Now find the speed of the ball immediately after impact.

Using conservation of mechanical energy from B to C

$$\tfrac{1}{2} mv^2 = mg(1.6) \qquad \Rightarrow \qquad v^2 = 31.36 \qquad \Rightarrow \qquad v = 5.6$$

$\therefore$ just after impact the speed of the ball is $5.6\,\mathrm{m\,s^{-1}}$.

(a) Using the law of restitution,

$$5.6 = 7e \qquad \Rightarrow \qquad e = 0.8$$

(b) The loss in mechanical energy due to the impact is the difference in kinetic energy immediately before and after impact.

$$\text{Loss in mechanical energy} = \tfrac{1}{2}(0.15)(7^2) - \tfrac{1}{2}(0.15)(5.6)^2 \text{ joules}$$

$$= 1.32\,\mathrm{J} \quad (3\text{ sf})$$

This loss may have been converted into sound energy at impact.

Kinetic energy can also be converted into heat at impact.

EXERCISE 5b

In questions 1 to 4 a particle moves directly towards a fixed plane surface, strikes it and rebounds. The coefficient of restitution is e.

1. If $e = \tfrac{1}{3}$,   find

    (a) the speed after impact

    (b) the impulse exerted by the particle on the plane.

2. If $e = \tfrac{2}{3}$,   find

    (a) the speed after impact

    (b) the loss of kinetic energy.

3. Given that $e = 0$,   find

    (a) the speed after impact

    (b) the impulse exerted by the particle on the plane

    (c) the loss of kinetic energy.

4. Given that $e = 1$,   find

    (a) the speed after impact

    (b) the impulse exerted by the particle on the plane

    (c) the loss of kinetic energy.

5. A small ball, of mass 0.02 kg, is dropped from a height of 0.4 m onto horizontal ground. The impact causes the ball to lose 75% of its mechanical energy. Find

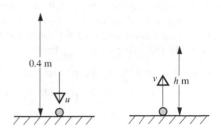

   (a) its speed just before impact

   (b) its speed just after impact

   (c) the coefficient of restitution

   (d) the height to which it rebounds.

6. A sphere of mass 2 kg falls from rest from a height of 10 m above an elastic horizontal plane. Find the height to which the sphere will rise again after its first bounce, if the coefficient of restitution is $\frac{1}{2}$. Find also the impulse exerted by the sphere on the plane.

   In each question from 7 to 11 a particle, travelling horizontally with velocity **u**, strikes a vertical wall at right angles and rebounds with velocity **v**. The coefficient of restitution is e. Draw a diagram to illustrate each question before solving it.

7. **u** = 8**i**, e = 0.75. Find **v**.

8. **u** = −5**j**,   **v** = 2**j**. Find e.

9. **v** = 4**i**, e = 0.2. Find **u**.

10. **u** = −6**i** + 10**j**, e = 0.5. Find **v**.

11. **u** = 6**i** − 8**j**,   **v** = −3**i** + 4**j**. Find e.

12. A small sphere is dropped onto a horizontal plane from a height of 20 m. The coefficient of restitution between the sphere and the plane is $\frac{1}{2}$.

   (a) Find the height to which the particle rises after each of the first, second and third impacts.

   (b) Show that the heights found in (a) are in geometric progression.

   (c) Find the total distance travelled by the sphere up to the fourth impact.

   (Use g = 10.)

*13. An ice-hockey player is skating up the rink at a constant speed of 4 m s⁻¹. When he is 10 m from the end he strikes the puck, giving it a speed of 20 m s⁻¹. The puck hits the end boards and rebounds directly towards him so that he receives it back one second after hitting it. Stating any simplifying assumptions which you make, construct and use a mathematical model to find the coefficient of restitution between the puck and the end boards.

*14. A light inextensible string AB, of length a, has the end A fixed to a vertical wall. The end B is attached to a particle which is drawn away from the wall until the string makes an angle of $60°$ with the wall. The particle is then released from rest. If the value of e is $\frac{3}{4}$, find

(a) the velocity of the particle just before impact with the wall

(b) the velocity of the particle just after impact

(c) the vertical distance through which the particle travels before it next comes to instantaneous rest.

*15. Some manufacturers are making a table game in which a small disc is projected by a spring mechanism to slide over a horizontal surface, hit a vertical back-board and rebound into areas marked with scores. The mass of a disc is 0.1 kg. The spring mechanism allows the disc to hit the back-board with speeds between $0.5\,\text{m s}^{-1}$ and $3\,\text{m s}^{-1}$. The coefficient of friction between a disc and the surface is 0.2. Dimensions are shown on the diagram. Model the game by treating the disc as a particle and assume that it strikes the back-board at right-angles.

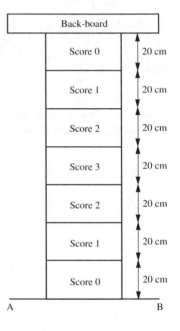

(a) In choosing material for the back-board it is required that the coefficient of restitution, e, should be such that the disc cannot rebound past the line AB. Find the maximum possible value of e for this requirement to be satisfied.

(b) Given that $e = 0.75$, find the range of speeds with which the disc can hit the back-board when the score is 3.

COLLISION OF TWO OBJECTS BOTH FREE TO MOVE

The principle of conservation of linear momentum can be applied to the direct impact of particles both of which are free to move, as this is a case where the pair of impulses at impact cause equal and opposite changes in momentum and so have no overall effect on the total momentum of the system.

For this situation therefore, in addition to the law of restitution, we can also use conservation of momentum.

Because we will now be dealing with more than one moving particle, it is particularly important to· choose a positive direction when dealing with the momentum of the system.

Examples 5c

1. A particle A, of mass 0.1 kg, is moving with velocity $2\,\mathrm{m\,s}^{-1}$ directly towards another particle B, of mass 0.2 kg, which is at rest. Both particles are on a smooth horizontal table. If the coefficient of restitution between A and B is 0.8, find the velocity (i.e. the speed and direction of motion) of each particle after their collision.

0.1 kg
A $\text{O}\!\!\longrightarrow\!\!\triangleright\, 2\,\mathrm{m\,s}^{-1}$

0.2 kg
O B

approach speed = 2 m s^{-1}

$J \longleftarrow\!\!\!\text{OO}\!\!\longrightarrow\! J$

$\longrightarrow$ positive (+ve) direction

0.1 kg
$\text{O}\!\!\longrightarrow\!\!\triangleright u$

0.2 kg
$\text{O}\!\!\longrightarrow\!\!\triangleright v$

separation speed = $v - u$

We know that B moves in the +ve direction after impact, as B is at rest when an impulse acts on it in that direction. However, we are not sure which way A will move after impact. We have marked its velocity as +ve and the sign we get for u will determine whether or not this is the correct direction.

Using conservation of momentum $\rightarrow$

$$0.1 \times 2 = 0.1u + 0.2v$$

$\Rightarrow \qquad\qquad u + 2v = 2$ [1]

Using the law of restitution

$$v - u = 0.8 \times 2 \qquad\qquad [2]$$

[1] + [2] gives $\qquad 3v = 3.6 \qquad \Rightarrow \qquad v = 1.2$

Then from [1] $\qquad u = 2 - 2.4 = -0.4$

The minus sign shows that A is, in fact, moving in the −ve direction after impact.

$\therefore$ after impact, the velocity of B is $1.2\,\mathrm{m\,s}^{-1}$ in the +ve direction

and $\qquad\qquad$ the velocity of A is $0.4\,\mathrm{m\,s}^{-1}$ in the −ve direction.

Note that it is easier to combine the two equations if equation [1] is arranged with u and v on the LHS.

At the moment of impact two equal and opposite impulses act, one on each of the particles. So if the magnitude of each impulse is required, we must consider the change in momentum of *only one* of the particles.

In the example above, for example, we would find the value of J by considering the motion of B (easier than A as B has no initial momentum),

i.e. $\xrightarrow{+} \qquad J = 0.2v - 0$

2. A particle P of mass 1 kg, moving with speed $4\,\mathrm{m\,s^{-1}}$, collides directly with another particle Q of mass 2 kg whose speed is $2\,\mathrm{m\,s^{-1}}$. The coefficient of restitution for these particles is 0.5. Find the velocity of each particle just after the collision and the magnitude of the impulses acting on impact, if before impact the particles are moving in   (a) opposite directions   (b) the same direction.

(a)

approach speed = 4 + 2

(+ve) direction

separation speed = $v - u$

Using conservation of momentum $\rightarrow$

$$1 \times 4 - 2 \times 2 = 1 \times u + 2 \times v$$

$\Rightarrow$ $u + 2v = 0$ [1]

Using the law of restitution

$$v - u = 0.5(4 + 2)$$ [2]

[1] + [2] gives $3v = 3$ $\Rightarrow$ $v = 1$ and $u = -2$

i.e. after impact P's speed is $2\,\mathrm{m\,s^{-1}}$ and Q's speed is $1\,\mathrm{m\,s^{-1}}$.

The direction of motion of both particles is reversed by the impact.

Now consider the impulse that acts on Q and the change in momentum it produces.

Impulse = final momentum − initial momentum

$\Rightarrow$ $J = 2 \times 1 - 2 \times (-2) = 6$

The magnitude of each impulse is 6 N s.

(b)

approach speed = 4 − 2

(+ve) direction

separation speed = $v - u$

Using conservation of momentum $\rightarrow$

$$4 \times 1 + 2 \times 2 = 1 \times u + 2 \times v \qquad \Rightarrow \qquad u + 2v = 8 \qquad\qquad [1]$$

Using the law of restitution

$$v - u = 0.5\,(4 - 2) \qquad\qquad\qquad\qquad\qquad\qquad [2]$$

$[1] + [2]$ gives $\qquad 3v = 9 \qquad \Rightarrow \qquad v = 3 \quad$ and $\quad u = 2$

i.e. after impact P's speed is $2\,\mathrm{m\,s}^{-1}$ and Q's speed is $3\,\mathrm{m\,s}^{-1}$, both particles travelling in the same direction as before impact.

Consider the impulse that acts on Q and the change in momentum it produces.

If $J\,\mathrm{N\,s}$ is the impulse that acts on Q then

$$J = 2 \times 3 - 2 \times 2 = 2$$

The magnitude of each impulse is $2\,\mathrm{N\,s}$.

3. A sphere of mass 1.5 kg and speed $2\,\mathrm{m\,s}^{-1}$ collides head-on with an identical sphere of mass 0.5 kg and speed $1\,\mathrm{m\,s}^{-1}$. As a result of the impact the direction of motion of the lighter sphere is reversed and its speed becomes $3.5\,\mathrm{m\,s}^{-1}$. Find

(a) the speed of the heavier sphere after impact

(b) the coefficient of restitution between the spheres

(c) the loss in kinetic energy due to the impact.

(d) What can you deduce from the answers to (b) and (c)?

(e) State an assumption that has been made.

(a) Using conservation of momentum $\rightarrow$

$$1.5 \times 2 - 0.5 \times 1 = 1.5u + 0.5 \times 3.5 \qquad \Rightarrow \qquad u = 0.5$$

The speed of the heavier sphere is $0.5\,\mathrm{m\,s}^{-1}$.

(b) Using the law of restitution, $3.5 - 0.5 = 3e$

$\Rightarrow \qquad\qquad e = 1$

(c)   KE before impact $= \frac{1}{2}(1.5)(2)^2 + \frac{1}{2}(0.5)(1)^2 = 3.25$

KE after impact $= \frac{1}{2}(1.5)(0.5)^2 + \frac{1}{2}(0.5)(3.5)^2 = 3.25$

$\therefore$ there is no loss in energy because of the impact.

(d)   From (b) we see that in this problem the spheres are perfectly elastic and, from (c), that no kinetic energy is lost. We deduce that, in general, a perfectly elastic impact causes no loss in kinetic energy. It follows that there is no 'bang' when a perfectly elastic collision occurs.

(e)   It is assumed that the spheres can be modelled as particles and do not distort on collision. This requires that the spheres are identical in size. If they were not, the impulses would not act in the direction of motion and so would affect the momentum in a perpendicular direction also.

EXERCISE 5c

In each question from 1 to 6, the diagram shows the velocities of two particles, moving on a smooth horizontal surface, just before and just after they collide.

1. The coefficient of restitution is $\frac{1}{3}$.

Just before impact   2 kg ○ ▷10 m s^{-1} 1 kg ○ ▷4 m s^{-1}

At impact J N s ◁——— ∞ ———▷ J N s

Just after impact   ○ ▷u m s^{-1} ○ ▷v m s^{-1}

Find the values of u, v, and J.

2.

Just before impact   1 kg ○ ▷16 m s^{-1} 2 kg ○ ▷3 m s^{-1}

Just after impact   ○ (at rest)   ○ ▷v m s^{-1}

Find   (a)   the value of v (b)   the coefficient of restitution.

Find also the kinetic energy lost at impact.

3. The coefficient of restitution is $\frac{1}{2}$.

Just before impact   1 kg ○ ▷8 m s^{-1} 2 m s^{-1} ◁○ 2 kg

Just after impact u m s^{-1} ◁○   ○ ▷v m s^{-1}

Find   (a) u (b) v (c)   the kinetic energy lost at impact.

4.

	3 kg		2 kg
Just before impact	O $\longrightarrow$ 1 m s^{-1}	5 m s^{-1} $\longleftarrow$	O
At impact	J N s $\longleftarrow$ OO	$\longrightarrow J$ N s	
Just after impact	u m s^{-1} $\longleftarrow$ O	O $\longrightarrow$ 1 m s^{-1}	

Find   (a)   the coefficient of restitution   (b)   the kinetic energy lost at impact.

5. A sphere A of mass 0.1 kg is moving with speed 5 m s^{-1} when it collides directly with a stationary sphere B. If A is brought to rest by the impact and $e = \frac{1}{2}$, find the mass of B, its speed just after impact and the magnitude of the instantaneous impulses.

6. A car of mass 1000 kg is waiting at rest at traffic lights and its driver has neglected to put the handbrake on. Another car, of mass 800 kg, approaches from behind and, braking too late, is travelling at 25 km h^{-1} when it hits the stationary car, giving it an initial speed of 20 km h^{-1}. Stating any assumptions which you make, estimate the coefficient of restitution between the cars.

7. An electron collides directly with a hydrogen atom which is initially at rest. The collision is perfectly elastic and the mass of the hydrogen atom is 1840 times the mass of the electron. What percentage of the electron's initial kinetic energy is transferred to the hydrogen atom?

8. Particles A and B both have mass m and are moving in the same direction along a line, A with speed $3u$ and B with speed u. They collide and after the impact they move in the same direction, A with speed u and B with speed ku. The coefficient of restitution is e.

Just before impact A	$\overset{m}{O} \longrightarrow 3u$	B $\overset{m}{O} \longrightarrow u$	
Just after impact		A $\overset{m}{O} \longrightarrow u$	B $\overset{m}{O} \longrightarrow ku$

(a)   Show that $e = \dfrac{k-1}{2}$.

(b)   Deduce that $1 \leqslant k \leqslant 3$.

(c)   Find the loss of kinetic energy in terms of m, k and u.

9. A sphere of mass 0.2 kg, moving at $10\,\mathrm{m\,s^{-1}}$, collides directly with another sphere, of mass 0.5 kg moving in the same direction at $8\,\mathrm{m\,s^{-1}}$. Their speeds after the collision are u and v respectively and the coefficient of restitution is e.

(a) Find expressions for u and v in terms of e.

(b) Show that $\frac{50}{7} \leqslant u \leqslant \frac{60}{7}$ and find similar inequalities for v.

10. A particle A, of mass m moving with a speed u collides directly with a particle B, of mass km, which is initially at rest. The direction of motion of A is reversed by the impact. The coefficient of restitution is $\frac{1}{3}$.

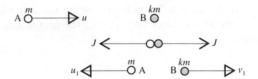

Giving answers in terms of m, k and u,

(a) find the speed of A after the impact.

(b) deduce that $k > 3$

(c) find the speed of B after the impact.

(d) find the impulse exerted by A on B.

11. A small sphere A, of mass m, moving with speed ku (where $k > 1$) collides directly with a small sphere B, of mass $3m$, which is moving with speed u in the same direction. The coefficient of restitution is $\frac{1}{2}$. Find in terms of k and u, an expression for the velocity of A after the impact and hence show that if the direction of motion of A is unchanged by the impact then $k < 9$.

*12. Two particles P and Q, of masses m and $3m$ respectively, are connected by a light elastic string of natural length l and modulus of elasticity $3mg$. They are held at rest on a smooth horizontal plane, with the string stretched to a length $5l$, and released from rest.

(a) By considering momentum, show that if V_P and V_Q are their speeds towards each other at any time between being released and colliding, then $V_P = kV_Q$, where k is a constant, and state the value of k.

(b) By considering energy find, in terms of g and l, their speeds just before collision.

Given that they adhere to each other after the collision

(c) find their common velocity just after the impact

(d) find, in terms of m, g and l, the impulse each exerts on the other at impact.

MULTIPLE IMPACTS

Sometimes a collision between two objects leads to further collisions either with another moveable object or with a fixed surface. In such cases each individual impact can be dealt with by the methods already described. It is best to solve one collision completely before starting on the next one, and to begin again with a new set of diagrams. The positive direction can be chosen afresh for each collision – it need not be the same throughout.

Examples 5d

1. An unfortunate snooker player miscues when he strikes the cue ball, giving it a speed of $8\,m\,s^{-1}$, so that it hits the brown ball directly and sets it moving towards, and perpendicular to, the cushion. The direction of motion of the cue ball is unaltered by the impact. After hitting the cushion the brown ball collides head-on with the cue ball. If the snooker balls are modelled as particles, several assumptions have to be made. Name any three of them.

   Given that the coefficient of restitution between the balls is 0.9 and between the ball and the cushion is 0.8, find the speed of the cue ball after its second collision.

   The cue ball is not given any spin; the table and the cushion are smooth; the balls are identical in size (so that the impulses are horizontal). Also, as the masses of the balls are not given, we must assume them to be equal.

   1st impact (between the balls)

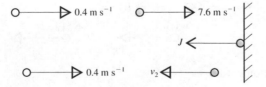

   Using conservation of momentum →

   $$8m = mu_1 + mv_1 \qquad \Rightarrow \qquad 8 = u_1 + v_1$$

   Using the law of restitution

   $$0.9 \times 8 = v_1 - u_1 \qquad \Rightarrow \qquad 7.2 = v_1 - u_1$$

   Adding gives $2v_1 = 15.2 \qquad \Rightarrow \qquad v_1 = 7.6 \quad\text{and}\quad u_1 = 0.4$

   2nd impact (with cushion)

   This is an external impact so momentum is not conserved.

Using the law of restitution,

$$0.8 \times 7.6 = v_2 \qquad \Rightarrow \qquad v_2 = 6.08$$

3rd impact (between the balls)

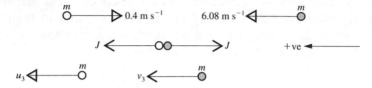

Using conservation of momentum ←

$$6.08m - 0.4m = mu_3 + mv_3 \qquad \Rightarrow \qquad 5.68 = u_3 + v_3$$

Using the law of restitution

$$0.9 \times 6.48 = u_3 - v_3 \qquad \Rightarrow \qquad 5.832 = u_3 - v_3$$

Adding gives $\quad 2u_3 = 11.512 \quad \Rightarrow u_3 = 5.76 \quad (\text{3 sf})$

The speed of the cue ball is $5.76\,\text{m s}^{-1}$ (3 sf).

Note that we used u for the speed of one ball and v for the other ball at every impact, then a suffix denotes the impact that has just taken place, e.g. u_1 is the speed of the cue ball after the first impact (u_2 does not appear as the cue ball was not involved in the second collision). This notation is particularly helpful when the number of impacts increases.

2. **A ball A of mass 0.3 kg moves on a smooth horizontal plane towards a ball B, of mass 0.2 kg which is at rest, and strikes it directly with speed $1\,\text{m s}^{-1}$. Ball B then moves with constant speed towards a third ball C, of mass 0.1 kg, which is lying at rest on B's line of motion. If the coefficient of restitution between A and B is 0.5 and that between B and C is 0.7, find the speed of each ball when no more collisions can take place between them.**

As several objects and several impacts are involved, we will denote the speeds of A, B and C by u, v and w, and use the suffixes 1, 2, 3... to denote the number of the collision that has just taken place. It is sometimes helpful to use u_0, etc. for the speeds before any impact has occurred.

1st collision (between A and B)

Conservation of momentum →

$$0.3 \times 1 = 0.3u_1 + 0.2v_1 \qquad \Rightarrow \qquad 3 = 3u_1 + 2v_1 \qquad\qquad [1]$$

Law of restitution

$$0.5 \times 1 = v_1 - u_1 \qquad \Rightarrow \qquad 0.5 = v_1 - u_1 \qquad\qquad [2]$$

$[1] + 3 \times [2]$ gives $\quad 4.5 = 5v_1 \qquad \Rightarrow \qquad v_1 = 0.9$ and $u_1 = 0.4$

2nd collision (between B and C)

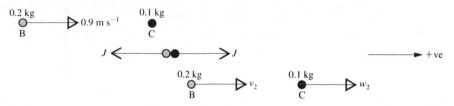

Conservation of momentum $\rightarrow$

$$0.2 \times 0.9 = 0.2v_2 + 0.1w_2 \qquad \Rightarrow \qquad 1.8 = 2v_2 + w_2 \qquad\qquad [3]$$

Law of restitution

$$0.7 \times 0.9 = w_2 - v_2 \qquad \Rightarrow \qquad 0.63 = w_2 - v_2 \qquad\qquad [4]$$

$[3] - [4]$ gives $\quad 1.17 = 3v_2 \qquad \Rightarrow \qquad v_2 = 0.39$ and $w_2 = 1.02$

After the second impact the velocities of the balls are:

A $\quad\longrightarrow$ 0.4 m s^{-1} $\qquad\qquad$ B $\quad\longrightarrow$ 0.39 m s^{-1} $\qquad\qquad$ C $\quad\longrightarrow$ 1.02 m s^{-1}

A's speed is greater than B's so A will catch up with B and collide again.

3rd collision (between A and B)

0.3 kg $\quad\longrightarrow$ 0.4 m s^{-1} $\qquad$ 0.2 kg $\quad\longrightarrow$ 0.39 m s^{-1}
A $\qquad\qquad\qquad\qquad\qquad$ B

$$J \longleftarrow \qquad\bigcirc\!\bullet\qquad \longrightarrow J \qquad\qquad\qquad \longrightarrow +\text{ve}$$

0.3 kg $\quad\longrightarrow u_3$ $\qquad$ 0.2 kg $\quad\longrightarrow v_3$
A $\qquad\qquad\qquad\qquad$ B

Conservation of momentum $\rightarrow$

$$0.3 \times 0.4 + 0.2 \times 0.39 = 0.3u_3 + 0.2v_3 \qquad \Rightarrow \qquad 1.98 = 0.3u_3 + 0.2v_3 \qquad [5]$$

Law of restitution

$$0.5 \times 0.01 = v_3 - u_3 \qquad\qquad\qquad\qquad\qquad\qquad [6]$$

$0.3 \times [6] + [5]$ gives $\quad 1.995 = 5v_3 \qquad \Rightarrow \qquad v_3 = 0.399$ and $u_3 = 0.394$

The velocities now are:

A $\quad\longrightarrow$ 0.394 m s^{-1} $\qquad\qquad$ B $\quad\longrightarrow$ 0.399 m s^{-1} $\qquad\qquad$ C $\quad\longrightarrow$ 1.02 m s^{-1}

B's speed is greater than A's but less than C's so no further impacts can occur.

The final speeds are: $\quad$ A: $0.394\,\text{m s}^{-1}$ $\qquad$ B: $0.399\,\text{m s}^{-1}$ $\qquad$ C: $1.02\,\text{m s}^{-1}$

EXERCISE 5d

In questions 1 to 3, two particles, A and B, lie on a smooth horizontal plane in a line that is perpendicular to a vertical wall. Initially B is at rest when A strikes it directly. B then goes on to strike the wall and rebounds.

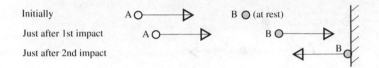

1. The mass of A is 1 kg and it strikes B with speed $4\,\text{m s}^{-1}$. B's mass is 8 kg.

Given that the coefficient of restitution at each impact is $\frac{1}{2}$, find u_1, v_1 and v_2. Explain why there is no further collision.

2. The masses of A and B are 2 kg and 1 kg respectively. A strikes B with speed $9\,\text{m s}^{-1}$. The coefficient of restitution between A and B is 1, and that between B and the wall is $\frac{1}{2}$.

   (a) Find the values of u_1 and v_1.

   (b) Show that there is a second collision between A and B and find their velocities after it. Is there a further collision?

3. Particle A, of mass 1 kg, strikes B, also of mass 1 kg, at $9\,\mathrm{m\,s^{-1}}$. The coefficient of restitution between A and B is $\frac{1}{3}$, and that between B and the wall is $\frac{1}{2}$.

Just before the first impact   1 kg   1 kg

$$\text{A} \!\!\!\bigcirc \!\!\longrightarrow\!\! 9\,\mathrm{m\,s^{-1}} \quad \text{B}\bigcirc\ (\text{at rest})$$

Find the velocities of A and B after each impact until no further collisions can occur.

In questions 4 to 7 three small smooth spheres, all with the same radius, lie in a straight line on a smooth horizontal plane. A is projected directly towards B which is at rest and after that collision B goes on to collide directly with C, also at rest.

4. The masses of A, B and C are 1 kg, 2 kg and 1 kg respectively. The coefficient of restitution between A and B is 1 and that between B and C is $\frac{1}{2}$.

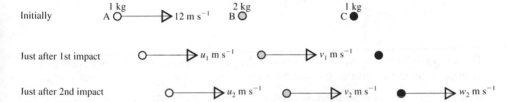

Find the velocities of each sphere after each possible impact.

5. The situation just before the first collision is shown in the diagram.

    2 kg   1 kg   2 kg

$$\text{A}\!\!\!\bigcirc\!\!\longrightarrow\!\! 8\,\mathrm{m\,s^{-1}} \quad\quad \text{B}\bigcirc\ (\text{at rest}) \quad\quad \text{C}\bullet\ (\text{at rest})$$

At each impact the coefficient of restitution is $\frac{1}{2}$. Find the velocities of A, B and C after each of the collisions that can occur.

6. The masses of A, B and C are $m\,\mathrm{kg}$, $m\,\mathrm{kg}$ and $km\,\mathrm{kg}$ respectively and A's initial speed is $u\,\mathrm{m\,s^{-1}}$. At each impact the coefficient of restitution is 1.

Just before the first impact $m\,\mathrm{kg}$ $m\,\mathrm{kg}$ $km\,\mathrm{kg}$

$$\text{A}\!\!\!\bigcirc\!\!\longrightarrow\!\! u\,\mathrm{m\,s^{-1}} \quad\quad \text{B}\bigcirc\ (\text{at rest}) \quad\quad \text{C}\bullet\ (\text{at rest})$$

(a) Find, in terms of u, the velocities of A and B after the first collision.

(b) Find, in terms of u and k, the velocities of B and C after the second collision.

(c) State, with reasons, whether there will be further collisions if
    (i) $k > 1$ (ii) $k < 1$.

7. A sphere A, of mass m_1, and velocity u, collides with a stationary sphere B of mass m_2. If sphere A is brought to rest by the collision, find

   (a) the velocity of B after impact

   (b) the coefficient of restitution.

   Sphere B now collides with a stationary sphere C and is brought to rest. Assuming the same coefficient of restitution between B and C as between A and B, find the mass of sphere C.

8. A ball game at a fair consists of a long groove in which there are two heavy balls, B with mass $4\,kg$ and C of mass $3\,kg$. A competitor bowls another ball A, of mass $3\,kg$, along the groove to strike B which, in turn, collides with C. The coefficient of restitution between A and B is $\frac{2}{3}$ and that between B and C is $\frac{1}{2}$. At the end of the groove there is an end-stop.

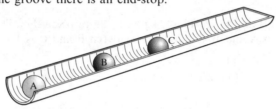

   The winner is the competitor who gives C the highest speed. (This is determined by measuring the time between the second impact and C's collision with the end-stop.)

   By modelling the balls as particles and the groove as smooth, find the speed given to C by a competitor who bowls A off at $10\,m\,s^{-1}$.

*9. A cunning competitor at the game described in question 10, checks that, to beat the current best score, he must give C a speed of $7\,m\,s^{-1}$. Find the speed at which he should project A to achieve this.

*10. Two beads, P of mass $0.01\,kg$ and Q of mass $0.02\,kg$, are threaded on to a smooth wire which is in the form of a circle in a horizontal plane. Initially Q is at rest and P strikes it with speed $0.75\,m\,s^{-1}$. After this impact they move in opposite directions round the circle. The coefficient of restitution is e.

   (a) Find their speeds after this first impact, in terms of e.

   (b) Given that they collide again after Q has rotated through $270°$ from its initial position, find the value of e and hence the values of the speeds after the first impact.

   (c) Find the velocities after the second impact.

*11. A small sphere is dropped on to a horizontal plane from a height h. The coefficient of restitution between the sphere and the plane is e.

   (a) Find, in terms of h and e, the height to which the particle rises after each of the first, second and third impacts.

   (b) Show that the heights found in (a) are in geometric progression.

   (c) Deduce the total distance travelled by the sphere before it comes to rest.

CHAPTER 6

CENTRE OF MASS

DEFINITIONS AND PROPERTIES

A summary follows of the definitions and methods given in Module E.

- The centre of mass of a body is the point about which the mass is evenly distributed. It follows that, for a uniform body, the centre of mass, G, lies on any axis of symmetry in the body. In normal situations this is also the point through which the weight of the body acts, i.e. the centre of gravity of the body.

- If the body comprises a number of sections, the sum of the moments of the weights of these sections about any axis is equal to the moment about the same axis of the weight of the whole body.

  In order to use this property to find G $(\bar{x}, \bar{y})$, the mass m_n, and the centre of mass (x_n, y_n), of each of the sections must be known. Then the centre of mass is given by $\Sigma m_n x_n = \bar{x} \Sigma m_n$ and $\Sigma m_n y_n = \bar{y} \Sigma m_n$

  Remember that this calculation can often be clearly set out in a table.

- The positions of the centres of mass of some simple uniform plane laminas are obvious.

A square or rectangular lamina has its centre of mass at the intersection of the diagonals.

The centre of mass of a uniform triangular lamina lies on any median, one-third of the way from the base to the vertex.

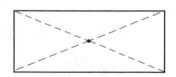

For a right-angled triangle this point is at a distance from the right angle equal to one third of the length of each adjacent side.

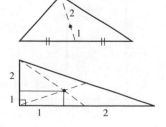

121

Check that you remember the background ideas by working through the following exercise.

EXERCISE 6a

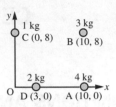

1. Find the centre of mass of the four particles shown in the diagram.

2. The particles defined in question 1 are now attached to a uniform rectangular lamina of mass 6 kg. Find the centre of mass of the complete system of lamina and particles.

3. A 50 cm length of uniform wire is bent into a square shape as shown in the diagram.
   A double thickness of wire forms the side AB.
   Find the distances of the centre of mass of the wire shape from AB and BC.

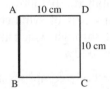

4. The mass of the earth is 5.976×10^{24} kg and that of the moon is 7.350×10^{22} kg. The distance between the centres at perigee (when they are closest) is 356 400 km. Find the distance of their centre of mass from the centre of the earth at this time.

5. The sides of triangle ABC are such that   AB = 10 cm,   BC = 24 cm   and   AC = 26 cm.

   (a) Show that triangle ABC is right-angled.

   (b) Find the distances of the centre of mass from AB and BC if the triangle is a framework of thin uniform wire.

6. If the triangle described in question 5 is a uniform lamina, find the distance of its centre of mass from AB and BC.

7. Find the coordinates of the centre of mass of each uniform lamina.

(a)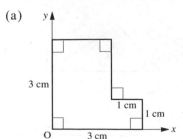

(b)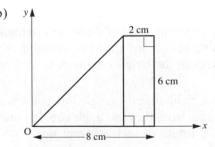

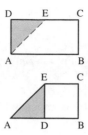

8. ABCD is a uniform thin rectangular metal
   plate. AB = 30 cm and BC = 15 cm.
   E is the mid point of CD. The plate is folded
   along the line AE so that corner D then lies at
   the mid point of AB. Find the distances of the
   centre of mass of the folded plate from AB
   and BC.

FURTHER PROBLEMS

In the revision exercise above all the compound objects involved no more than
two sections.

Now we look at some problems that involve finding the centre of mass of more
complex compound bodies.

Examples 6b

1. The diagram shows a uniform lamina, whose
   density is 1.2 kg per square metre, in the
   shape of a square ABCD, of side 2 m. A
   circular hole of radius 0.5 m is cut from it.
   The centre of the hole is distant 0.8 m from
   both AB and AD. A thin uniform
   strengthening strip is attached to the edge BC.
   Given that the mass of the strip is 1.5 kg,
   find the distances of the centre of mass
   of the lamina from AB and from AD.

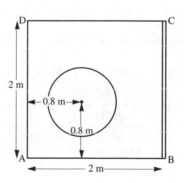

The mass of ABCD is $2^2 \times 1.2$ kg, i.e. 4.8 kg, and the mass removed for the hole
is $\pi(0.5)^2 \times 1.2$ kg, i.e. 0.9425 kg (using 4 sf during the working).

Portion	Mass	Distance of C of M		Mass × x	Mass × y
		x from AD	y from AB		
+ABCD	4.8	1	1	4.8	4.8
−Hole	0.9425	0.8	0.8	0.754	0.754
+Strip	1.5	2	1	3	1.5
Whole	5.358	$\bar{x}$	$\bar{y}$	$5.358\bar{x}$	$5.358\bar{y}$

Using $\Sigma mx = \bar{x}\Sigma m$ gives $(4.8 - 0.754 + 3) = 5.358\bar{x}$

$\Rightarrow$ $\bar{x} = 1.315$

Using $\Sigma my = \bar{y}\Sigma m$ gives $(4.8 - 0.754 + 1.5) = 5.358\bar{y}$

$\Rightarrow$ $\bar{y} = 1.035$

The centre of mass is 1.32 m from AD and 1.04 m from AB   (3 sf).

2. A 'modern art' wall plaque, shown in
the diagram, is made up from an
isosceles triangular uniform sheet of
copper, ABC, joined along the edge BC
to a uniform right-angled triangle of
aluminium, BCD, where
BC = BD = 0.6 m.   A lead motif
is attached at C.
The mass of the copper triangle is 1 kg
and that of the aluminium triangle is
0.4 kg.

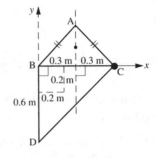

The artist intends the plaque to hang from A with BC horizontal, so its centre of
gravity must lie on the perpendicular bisector of BC. What must the mass of the
motif be for this to be achieved.

Only the x-coordinate of the centre of mass is needed and this is known to be 0.3.

Portion	Mass (m)	x at C of M	mx
ABC	1	0.3	0.3
BCD	0.4	0.2	0.08
Motif	M	0.6	$0.6M$
Plaque	$(1.4+M)$	0.3	$(1.4+M)(0.3)$

Using $\Sigma mx = \bar{x}\Sigma m$ gives $0.3 + 0.08 + 0.6M = (1.4 + M)(0.3)$

$\therefore$ $0.3M = 0.04$ $\Rightarrow$ $M = 0.133\ldots$

The mass of the motif should be 0.133 kg   (3 sf).

EXERCISE 6b

1. A uniform lamina has the shape of a square
ABCD, of side 30 cm, joined along the side CE to
an equilateral triangle CDE. Find the distance of
the centre of mass from AB.

2. The density of the lamina described in question 2 is $0.25\,\text{g/cm}^2$. The centre of mass is to be relocated on the line CE by attaching a particle of mass $m\,\text{g}$ at the point D. Find m.

3. The centre of mass of the lamina in question 2 is brought on to the line CE by making a different adjustment. This time the density of the triangular section is increased while the density of the square section remains at $0.25\,\text{g/cm}^2$. Find the density of the triangular section.

4. A circular lamina of radius $10\,\text{cm}$, with diameter AB, has a circular hole, of radius $r\,\text{cm}$, cut in it. The centre of the hole is at a point C on AB, where $\text{AC} = x\,\text{cm}$. It is required that after making this hole the centre of mass of the remainder should lie at a distance of $8\,\text{cm}$ from A. By finding the necessary value for x in each case, decide whether this can be done

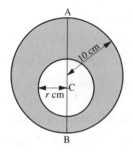

(a) if $r = 5\,\text{cm}$,

(b) if $r = 6\,\text{cm}$.

CENTRE OF MASS OF A RIGID BODY

So far we have been finding the centres of mass of laminas, i.e. two dimensional objects. Therefore only two coordinates were needed to locate the centre of mass.

Now we are going to consider rigid three-dimensional objects and the location of their centres of mass clearly requires three coordinates. In all the cases we deal with at this level, however, the body will have either an axis or a plane of symmetry so that at least one of the coordinates is obvious.

For example, a solid right cone has an axis of symmetry from the vertex to the centre of the base so its centre of mass, G, is somewhere on this line; it remains only to find one coordinate, i.e. the distance of G from the vertex (or the base).

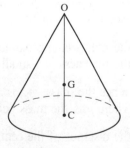

Whenever the centre of mass of a symmetrical solid object is to be found, the line of symmetry should be chosen as one of the axes of coordinates.

FINDING A CENTRE OF MASS BY INTEGRATION

When an object can be divided into a small number of parts, the mass and centre of mass of each part being known, the centre of mass of the whole object can be found by using $\Sigma m_n x_n = \bar{x} \Sigma m_n$ with similar expressions for $\bar{y}$ and $\bar{z}$ when the object is three-dimensional.

Some bodies cannot be divided up in this way, but can be divided into a very large number of very small parts whose masses and centres of mass are known. In cases like this, it may be possible to evaluate $\Sigma m_n x_n$ and Σm_n by using integration.

Suppose that we are asked to find the position of the centre of mass of a uniform solid right circular cone with height h and base radius a.

We will take the axis of symmetry as the x-axis. Therefore the centre of mass G lies on the x-axis and only the x-coordinate of G has to be found. For a reason that will become clear during the calculation we place the origin at the vertex.

The cone can be divided into thin slices, each parallel to the base and each being approximately a thin circular disc with its centre of mass on the x-axis.

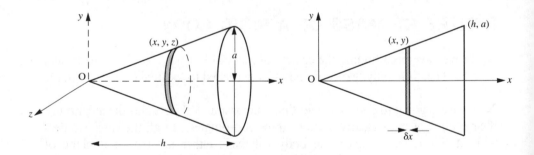

Considering one typical slice, called an *element*, we see that it is approximately a disc for which:

    the radius is approximately y,
    the thickness is a small increase in x, i.e. δx,

hence the mass, m_n, is approximately $(\pi y^2 \delta x)\rho$,
where ρ is the mass per unit volume,

the distance of the centre of mass from O is approximately x_n,
hence $m_n x_n$ is approximately $(\pi \rho y^2 \delta x)x$

Therefore $\Sigma m_n x_n = \Sigma \pi \rho y^2 x \delta x$

Now $\quad \underset{\delta x \to 0}{\text{limit}} \, \Sigma \pi \rho y^2 x \delta x = \int \pi \rho y^2 x \, \mathrm{d}x$

Therefore as $\quad \delta x \to 0 \qquad \Sigma m_n x_n \to \int \pi \rho y^2 x \, \mathrm{d}x$

The mass of the whole cone is $V\rho$ where V is the volume of the cone,

$\therefore \qquad \bar{x} \, \Sigma m_n = V \rho \, \bar{x}$

Then using $\quad \Sigma m_n x_n = \bar{x} \, \Sigma m_n \quad$ gives

$$\int \pi \rho y^2 x \, \mathrm{d}x = V \rho \, \bar{x} \qquad\qquad [1]$$

It is important to appreciate that equation [1] applies to *any* solid that is symmetrical about the x-axis and can be divided into disc-like elements.

For the cone in particular, $\quad V = \frac{1}{3} \pi a^2 h$, and from the similar triangles in the diagram we see that

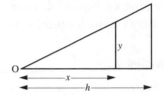

$$\frac{y}{x} = \frac{a}{h} \qquad \Rightarrow \qquad y = \frac{ax}{h}$$

This relationship is simple because the vertex is the origin of coordinates.

Now when equation [1] is applied to the cone,

the LHS becomes $\quad \displaystyle\int \pi \rho \left(\frac{ax}{h} \right)^2 x \, \mathrm{d}x$

and the RHS becomes $\quad \frac{1}{3} \pi a^2 h \rho \bar{x}.$

Now the value of x goes from 0 to h, hence [1] becomes

$$\frac{\pi \rho a^2}{h^2} \int_0^h x^3 \, \mathrm{d}x = \frac{1}{3} \pi a^2 h \rho \bar{x}$$

$\Rightarrow \qquad\qquad 3 \displaystyle\int_0^h x^3 \, \mathrm{d}x = h^3 \bar{x}$

$\Rightarrow \qquad\qquad \dfrac{3}{4} \Big[x^4 \Big]_0^h = h^3 \bar{x}$

Therefore $\qquad\qquad \bar{x} = \frac{3}{4} h$

i.e. **the centre of mass of a uniform solid cone is on the axis of symmetry and three-quarters of the way from the vertex to the base.**

Any uniform solid that is symmetrical about the x-axis, and for which all cross-sections are circular, is a *solid of revolution* and its centre of mass can be found by the process given above. It is only the total volume, and the boundaries of the integration, that differ from one such solid of revolution to another. So $\int \pi \rho y^2 x \ \mathrm{d}x = V \rho \bar{x}$ for *any* solid of revolution symmetrical about the x-axis,

i.e.
$$\int \pi y^2 x \ \mathbf{d}x = V \bar{x}$$

If a solid of revolution is symmetrical about the y-axis, it can be divided into slices with approximate radius x and thickness δy, leading to a similar result, i.e. $\int \pi x^2 y \ \mathrm{d}y = V \bar{y}$

Hence the location of the centre of mass on the y-axis can be found.

These results are quotable but it is advisable to look at a suitable element in each problem as a check on the expression to be integrated.

Examples 6c

1. **The diagram shows a quadrant of a circle of radius a and equation $x^2 + y^2 = a^2$. When this quadrant is rotated through a complete revolution about the x-axis a uniform hemisphere is generated.**

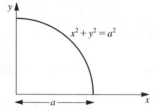

   Find the coordinates of the centre of mass of the hemisphere.

The hemisphere is symmetrical about the x-axis so the centre of mass, G, is on this axis. Therefore the y and z coordinates of G are both zero.

We will divide the hemisphere into slices parallel to the flat face.

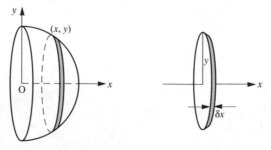

Considering one element as being almost a circular lamina then approximately: the radius is y, the thickness is δx and the distance of the element from O is x.

Hence the mass, m_n, of the element $\approx \pi y^2 \delta x \rho$.

The volume of the whole hemisphere is $\frac{2}{3} \pi a^3$.

Using the quotable result $\int y^2x\ dx = V\bar{x}$ gives

$$\int_0^a y^2x\ dx = \tfrac{2}{3}a^3\bar{x}$$

The integration cannot be carried out until y is expressed as a function of x.

From the equation of the circle, $y^2 = a^2 - x^2$

$$\therefore \qquad \int_0^a y^2x\ dx = \int_0^a x(a^2 - x^2)\ dx$$

$$= \left[\tfrac{1}{2}a^2x^2 - \tfrac{1}{4}x^4\right]_0^a = \tfrac{1}{4}a^4$$

$$\therefore \qquad \tfrac{2}{3}a^3\bar{x} = \tfrac{1}{4}a^4$$

$$\Rightarrow \qquad \bar{x} = \tfrac{3}{8}a$$

Therefore the centre of mass of the hemisphere is distant $\tfrac{3}{8}a$ from O on the radius of symmetry.

The coordinates of the centre of mass are $\left(\tfrac{3}{8}a, 0, 0\right)$.

2. **The area bounded by the y-axis, the line $y = 2$ and part of the curve with equation $y = 2x^2$ is rotated about the y-axis to give a uniform solid. Find the distance of the centre of mass of the solid from the origin O.**

The solid is symmetrical about the y-axis so the x and z coordinates of its centre of mass are both zero. The elemental slices are perpendicular to the y-axis.

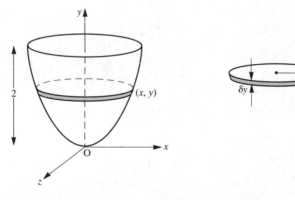

The approximate radius of the element is x, its thickness δy, therefore $m_n \approx \pi x^2 \delta y \rho$.

Quoting $\int \pi\rho x^2y\ dy = V\rho\bar{y}$ gives $\pi\int_0^2 x^2y\ dy = V\bar{y}$

Using $x^2 = \tfrac{1}{2}y$ the LHS becomes

$$\pi\int_0^2 \tfrac{1}{2}y^2\ dy = \pi\left[\tfrac{1}{6}y^3\right]_0^2 = \tfrac{4}{3}\pi$$

Now there is no formula for the volume of this solid of revolution, so the volume, V, must in this case be found by integration.

The volume of an elemental slice $\approx \pi x^2 \delta y$

$$\therefore \qquad V = \int_0^2 \pi x^2 \ \mathrm{d}y \qquad \left(\text{where} \ x^2 = \tfrac{1}{2}y\right)$$

$$= \pi \int_0^2 \tfrac{1}{2}y \ \mathrm{d}y = \pi \left[\tfrac{1}{4}y^2\right]_0^2$$

$$\Rightarrow \qquad V = \pi$$

Returning to $\qquad \pi \displaystyle\int_0^2 x^2 y \ \mathrm{d}y = V\bar{y} \qquad$ we have

$$\tfrac{4}{3}\pi = \pi\bar{y}$$

Therefore $\qquad\qquad\qquad\qquad \bar{y} = \tfrac{4}{3}$

THE CENTRE OF MASS OF A SEMICIRCLE

The integration method can also be used to find the centre of mass of a uniform semicircular lamina.

Consider such a lamina, bounded by the y-axis and part of the curve with equation $x^2 + y^2 = a^2$, divided into vertical strips as shown. The x-axis is a line of symmetry so the centre of mass lies on it.

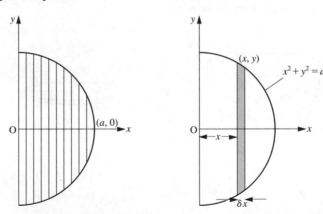

For one elemental strip of width δx,

the length is $2y$ so the area is approximately $2y\delta x$,
the mass is approximately $2y\rho\delta x$ where ρ is the mass per unit *area*,
the distance of the centre of mass from the y-axis is approximately x

Therefore $mx \approx (2y\rho\delta x)x$, i.e. $2xy\rho\delta x$

Now we can sum all the elements using $\Sigma mx = \bar{x}\Sigma m$ where

$$\sum_0^a mx = \sum_0^a 2xy\rho\delta x$$

As the width of the elemental strips approaches zero,

the limit as $\delta x \to 0$ of $\displaystyle\sum_0^a 2xy\rho\delta x = \int_0^a 2xy\rho\delta x$

and the limit as $\delta x \to 0$ of Σm is the mass of the semicircle, i.e. $\frac{1}{2}\pi a^2\rho$

Now $x^2 + y^2 = a^2 \quad\Rightarrow\quad y = \sqrt{(a^2 - x^2)}$

Therefore $\quad\quad 2\rho\displaystyle\int_0^a x\sqrt{(a^2 - x^2)}\ \mathrm{d}x = \frac{1}{2}\pi\rho a^2\bar{x}$

$\Rightarrow \quad\quad\quad\quad 2\left[-\frac{1}{3}(a^2 - x^2)^{3/2}\right]_0^a = \frac{1}{2}\pi a^2\bar{x}$

$\Rightarrow \quad\quad\quad\quad -\frac{2}{3}[0 - a^2] = \frac{1}{2}\pi a^2\bar{x}$

Hence $\quad \bar{x} = \dfrac{4a}{3\pi}$

i.e.   **the centre of mass of a uniform semicircular lamina**
                    is on the radius of symmetry
         and distant $\dfrac{4a}{3\pi}$ **from the centre of the plane edge.**

This result is quotable.

The centres of mass of other uniform laminas bounded by curves with known equations, can be found in a similar way but it is unlikely that a formula exists for the area of such a lamina. This means that Σm has to be found by integration.

Suppose, for example, that we want to find the centre of mass of a uniform lamina in the shape of the area in the first quadrant bounded by the x and y axes, and the curve $y = 9 - x^2$.

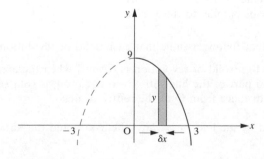

Using vertical elements gives $\quad \bar{x} \Sigma m = \Sigma xy\rho\delta x \quad$ as before.

As before, the r.h.s. becomes $\displaystyle \int_0^3 xy\rho\delta x = \int_0^3 x(9 - x^2)\rho \, dx$

but the l.h.s. cannot be evaluated by formula.

Instead we use $\quad \Sigma m = \Sigma y\rho\delta x$

$\Rightarrow \qquad$ the limit as $\delta x \to 0$ of $\displaystyle \sum_{x=0}^{x=3} m = \int_0^3 y\rho\delta x = \int_0^3 (9 - x^2)\rho \, dx$

Hence $\quad \bar{x} \displaystyle \int_0^3 (9 - x^2)\rho \, dx = \int_0^3 x(9 - x^2)\rho \, dx$

i.e. $\qquad \bar{x} \displaystyle \int_0^3 (9 - x^2) \, dx = \int_0^3 x(9 - x^2) \, dx$

So we see that two separate integrations have to be carried out before $\bar{x}$ can be found.

EXERCISE 6c

1. The shape of a uniform lamina is that of the area bounded by the x-axis, the curve $y = x^2$ and the line $x = 2$.

   (a) Find the area of the lamina

   (b) Find the x-coordinate of the centre of mass of the lamina.

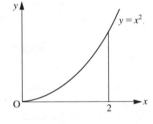

2. For a uniform lamina in the shape of the area in the first quadrant between the curve $y = x^2$ and the line $y = 4$, find

   (a) the area

   (b) the y-coordinate of the centre of mass (use elemental strips parallel to the x-axis).

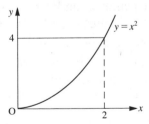

In the questions that follow, assume that each solid of revolution is uniform.

In questions 3 to 10 a solid of revolution is formed when the area bounded by the given lines and part of the line with given equation, is rotated about the x-axis. Find the distance from O of the centre of mass.

3. The area bounded by the line $y = 2x$, the x-axis and the lines $x = 2$ and $x = 4$.

4. The area between the line $y = ax$ and the x-axis, from $x = h$ to $x = 2h$.

5. The area in the first quadrant bounded by $y = x^2$, the x-axis and the line $x = 2$.

6. The area bounded by $y = x^2 + 2$, the x-axis, the y-axis and the line $x = 1$.

7. The area bounded by $y = \sqrt{x}$, $x = 1$, $x = 4$ and the x-axis.

8. The area bounded by $y = \sqrt{x+4}$, the x-axis and the lines $x = 0$ and $x = 3$.

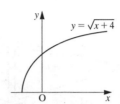

9. The area bounded by $y = e^x$, $x = 1$ and the x and y axes.

10. The area bounded by $y = \dfrac{1}{\sqrt{x+1}}$, the x-axis, the y-axis and the line $x = 5$.

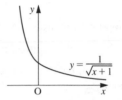

In questions 11 to 14 the area bounded by the given curves and lines is rotated about the y-axis to form a solid of revolution. Find the distance of the centre of mass from the origin.

11. The area in the first quadrant bounded by $y = \dfrac{1}{x^2}$, $y = 1$, $y = 5$ and the y-axis.

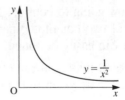

12. The area bounded by $x^2 + y^2 = 64$ and the positive x and y axes.

13. The area bounded by $y = 4 - x^2$, the x-axis and the y-axis.

14. The area in the first quadrant bounded by $x^2 + y^2 = 25$, $y = 3$, $y = 4$ and the y-axis.

15. A solid sphere of radius a is cut into two sections by making a plane cut at a distance $\frac{1}{2}a$ from the centre. Find, by integration, the distance of the centre of mass of the smaller of the two sections from the centre of the original sphere.

*16. The perpendicular height of a solid, right pyramid is h and the base is a square of side $2a$. Find the volume of the pyramid.

The vertex of the pyramid is taken as the origin O, the x-axis is the axis of symmetry and the position of the pyramid is shown in the diagram. The pyramid is divided into slices perpendicular to the x-axis.

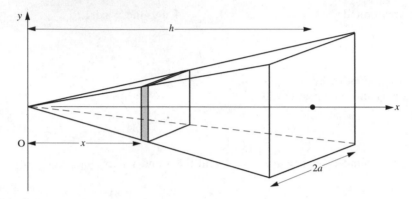

Considering an elemental slice distant x from O,

(a) by using similar figures, express the area of the slice in terms of x, a and h

(b) find the approximate mass, m, of the element

(c) find Σmx

(d) find the position of the centre of mass of the pyramid.

COMPOUND SOLIDS

We are now going to consider finding the centre of mass of a solid body that is made up of two (or more) parts, each of whose volume and centre of mass is known or can easily be found.

The centres of mass you can quote, in addition to the obvious cases of a cuboid and a sphere, are given in the table below.

Body	Position of G
Cylinder	Halfway up the centre line.
Hemisphere	On the radius of symmetry, $\frac{3}{8}$ of the way from the centre of the plane face.
Cone	On the axis of symmetry, $\frac{1}{4}$ of the perpendicular height above the base.
Pyramid	As for a cone.

The method we use when dealing with compound solids is the same as that used for compound laminas, i.e. the mass and location of centre of mass (C of M) of each part, and of the whole body, are tabulated, providing the data necessary to apply $\Sigma m_n x_n = \bar{x} \Sigma m_n$.

Examples 6d

1.

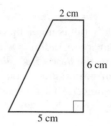

2 cm

6 cm

5 cm

The cross-section of a uniform solid prism of length l, is the trapezium shown in the diagram. Find the centre of mass of the prism.

The plane that divides the prism into equal halves is a plane of symmetry, so the centre of mass G lies in this plane which we will take as the xy plane.

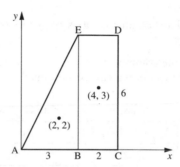

Taking ρ as the mass per unit volume we have:

Portion with cross-section	Mass m	Coords of C of M		mx	my
		x	y		
+ABE	$\frac{1}{2}(3)(6)l\rho$	2	2	$18\,l\rho$	$18\,l\rho$
+BCDE	$(2)(6)l\rho$	4	3	$48\,l\rho$	$36\,l\rho$
+ABCDE	$(9+12)l\rho$	$\bar{x}$	$\bar{y}$	$21\,l\rho\bar{x}$	$21\,l\rho\bar{y}$

Using $\Sigma m_n x_n = \bar{x}\,\Sigma m_n$ gives

$$18\,l\rho + 48\,l\rho = 21\,l\rho\bar{x} \quad \Rightarrow \quad \bar{x} = \tfrac{22}{7}$$

Using $\Sigma m_n y_n = \bar{y}\,\Sigma m_n$ gives

$$18\,l\rho + 36\,l\rho = 21\,l\rho\bar{y} \quad \Rightarrow \quad \bar{y} = \tfrac{18}{7}$$

Therefore G lies on the plane of symmetry at the point distant $\tfrac{22}{7}$ cm and $\tfrac{18}{7}$ cm from A in the directions of AC and CD respectively.

Note that when a solid has a *plane* of symmetry, finding the centre of mass amounts to finding the centre of mass of a lamina in the shape of the cross section.

2. **A child's toy is made up of a uniform solid hemisphere of radius a with its plane face fixed to the base of a solid right circular cone of base radius a and height $4a$. Find the distance of G, the centre of mass of the toy, from the centre of the common face.**

The hemisphere and cone have a common axis of symmetry so the centre of mass lies on it.

Taking O as the centre of the common face and the positive x-axis through the vertex of the cone, the following table can be compiled.

Section	Mass, m	x at C of M	mx
Cone	$\frac{1}{3}\pi a^2 (4a)\rho$	$+a$	$\frac{4}{3}\pi a^4 \rho$
Hemisphere	$\frac{2}{3}\pi a^3 \rho$	$-\frac{3}{8}a$	$-\frac{1}{4}\pi a^4 \rho$
Whole body	$2\pi\pi a^3 \rho$	$\bar{x}$	$2\pi a^3 \rho\bar{x}$

Using $\Sigma m_n x_n = \bar{x}\,\Sigma m_n$ gives

$$\frac{4}{3}\pi a^4 \rho + \left(-\frac{1}{4}\pi a^4 \rho\right) = 2\pi a^3 \rho\bar{x}$$

$\Rightarrow$ $\frac{13}{12}a = 2\bar{x}$

The positive sign for $\bar{x}$ shows that G is in the cone and not the sphere.

Therefore G is on the axis of symmetry of the cone, distant $\frac{13}{24}a$ from the common face.

3. **The diagram shows a uniform solid body formed from a cube of edge $4a$ with a cube of edge $2a$ removed from one corner. Find the position of the centre of mass of the solid.**

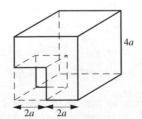

The most convenient choice of origin is the 'corner' that has been removed because this point is at the corner of both of the cubes we consider.

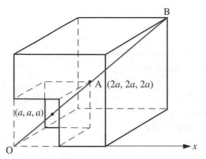

There is an axis of symmetry, passing through O, A and B. The centre of mass, G, is therefore on OAB and is equidistant from the three faces that meet at O. Therefore we need find only one of these distances.

Section	Mass, m	x at C of M	mx
+ Large cube (complete)	$(4a)^3 \rho$	$2a$	$128a^4 \rho$
− Small cube	$(2a)^3 \rho$	a	$8a^4 \rho$
Remaining body	$(4^3 - 2^3)a^3 \rho$	$\bar{x}$	$56a^3 \rho\bar{x}$

Using $\quad \Sigma m_n x_n = \bar{x}\, \Sigma m_n \quad$ gives

$$128a^4 \rho - 8a^4 \rho = 56a^3 \rho\bar{x}$$

$\Rightarrow \qquad\qquad\qquad 120a = 56\bar{x}$

$\Rightarrow \qquad\qquad\qquad \bar{x} = \frac{15}{7}a$

The centre of mass of the solid is distant $\frac{15}{7}a$ from the removed corner, along each of the edges.

Note that by using *the ratio of volumes of similar solids*, the table can be made even simpler. The ratio of the volumes, and therefore the masses, of the two cubes is $(4a)^3 : (2a)^3$, i.e. $8:1$.

Therefore, taking M as the mass of the removed cube, the table becomes:

Section	Mass m	x at C of M	mx
+ Large cube	$8M$	$2a$	$16aM$
− Small cube	M	a	aM
Remaining body	$7M$	$\bar{x}$	$7M\bar{x}$

giving $\qquad 16aM - aM = 7M\bar{x} \qquad \Rightarrow \qquad \bar{x} = \frac{15}{7}a$

For those readers who are comfortable with using similarity, this method, where appropriate, is obviously neater.

4. A uniform solid right cone is of height $12h$. The upper part of the cone is removed by a cut parallel to the base at a distance of $4h$ from the vertex, forming a frustum of a cone. Find the height above the centre of the base of the centre of mass of the frustum.

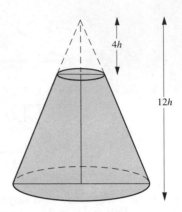

Although we are asked to locate G from the base, we will place O at the original vertex.

The removed cone and the original cone are similar, with linear a ratio of $4h:12h$, i.e. $1:3$. So the masses are in the ratio $1:27$. If M is the mass of the cone removed, the mass of the original cone is $27M$.

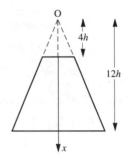

Section	Mass m	x at C of M	mx
+ Large cone	$27M$	$\frac{3}{4}(12h)$	$243Mh$
− Small cone	M	$\frac{3}{4}(4h)$	$3Mh$
Frustum	$26M$	$\bar{x}$	$26M\bar{x}$

Using $\Sigma m_n x_n = \bar{x}\,\Sigma m_n$ gives

$$243Mh - 3Mh = 26M\bar{x} \quad \Rightarrow \quad \bar{x} = \tfrac{240}{26}h = \tfrac{120}{13}h$$

Therefore the height of G above the base is $12h - \tfrac{120}{13}h = \tfrac{36}{13}h$

There are occasions when a compound object is made of sections which do not all have the same mass per unit volume. In these cases the separate values of the densities may be given or, instead, the ratio of the densities of the parts may be given.

5. A concrete bollard comprises a solid cylinder, of radius 8 cm and height 46 cm, surmounted by a cone of equal radius and height 24 cm. The weight per unit mass of the concrete from which the cylindrical section is made is twice that of the mix in the conical section. Find the height of the centre of mass of the bollard.

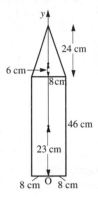

Let the density of the material in the cone be ρ so that the material in the cylinder is of density 2ρ.

Portion	Mass m	y at C of M	my
Cylinder	$\pi(8)^2(46)(2\rho)$	23	$92 \times 23(8)^2 \pi\rho$
Cone	$\frac{1}{3}\pi(8)^2(24)\rho$	$46+6$	$8 \times 52(8)^2 \pi\rho$
Bollard	$(92+8)(8)^2 \pi\rho$	$\bar{y}$	$100(8)^2 \pi\rho\,\bar{y}$

$$\Sigma my^2 = \bar{y}\,\Sigma m$$

$\Rightarrow \quad 92 \times 23(8)^2 \pi\rho + 8 \times 52(8)^2 \pi\rho = 100(8)^2 \pi\rho\bar{y}$

$\Rightarrow \quad\quad\quad\quad\quad \bar{y} = 25.32\ldots$

The centre of mass of the bollard is 25.3 cm from the base (3 sf)

Useful Points to Consider When Finding the Centre of Mass of a Solid Body

- If there is a plane of symmetry use it if possible as the xy plane (i.e. the plane containing the x and y axes.)

- If there is an axis of symmetry, use it as the x-axis (or the y-axis)

- Place O at a point from which the centre of mass of each section is easy to locate.

- If O is *inside* the solid, remember that some value(s) of x will be negative.

- If the body is made up of similar sections, the ratio of masses can be found from the ratio of volumes.

EXERCISE 6d

1. A stool has a top that is 30 cm square and 2 cm thick. At each corner there is a leg, 20 cm long and with a cross-section that is a 3 cm square. The top and legs are made from the same material. Find the distance from the top of the stool to the centre of mass of the stool.

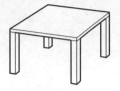

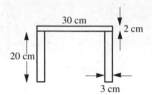

Questions 2 to 4 are about a uniform solid made by joining a right circular cone to a cylinder. Take the centre of the common face as origin and the x-axis along the line of symmetry as shown.

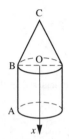

2. Find the distance of the centre of mass from the common face in each case.

(a)   (b)

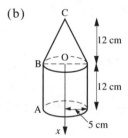

3. The centre of mass lies in the common face of the cone and the cylinder. If h is the height of the cone and H is the height of the cylinder, show that $h^2 = 6H^2$.

4. Given the dimensions in the diagram, find the distance of the centre of mass from the common face. Give your answer in terms of h.

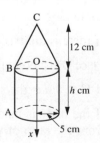

Questions 5 and 6 are about a frustum of a
uniform solid cone. (It is a good idea to take
the vertex of the original cone as origin).

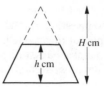

5. Given that $H = 24$ and $h = 12$, find the distance of the centre of mass
from the larger plane face of the frustum.

6. If $H = 3a$ and $h = a$, find the distance of the centre of mass from the
larger plane face of the frustrum.

Questions 7 and 8 are about the uniform solid,
shown in the diagram, made by joining the
plane face of a hemisphere to the plane face of
a cone.

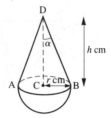

7. Find the distance of the centre of mass from the common plane face of the cone
and the hemisphere

   (a) if $r = 5$ and $h = 20$ (b) if $r = 15$ and $h = 10$.

8. The centre of mass lies in the common plane face of the cone and the hemisphere.
Find α.

9. A uniform solid is made by joining the plane face of a hemisphere of radius r to
one plane face of a cylinder of radius r and height h.

   (a) Find the distance of the centre of mass from the common plane face.

   (b) Find r in terms of h given that the centre of mass lies in the common plane
   face.

10. A solid cylinder, of height $2a$ and radius $2a$, has a
hemisphere removed from its upper plane end.
The hemisphere has radius a and its centre is on
the axis of the cylinder. Find the depth below the
top of the centre of mass of the remaining solid.

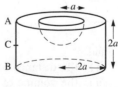

*11. A solid consists of a cone, of height 80 cm, which
has had a cone of the same radius and of height h
centimetres removed from its base. The centre of
mass of the remaining solid is 15 cm above the
base. Find the value of h.

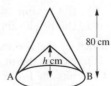

*12. An A level Maths student is planning a cycle tour. As a modelling exercise he decides to investigate which of two methods of carrying his luggage will have the lower centre of mass. He has a saddlebag, which he models as a cylinder, and two pairs of pannier bags, one pair larger than the other, which he models as prisms with a trapezium cross-section. He makes the following simplifying assumptions.

- The masses of the means of attachment are negligible.
- All the bags are fully packed to a uniform density.

The two methods he considers are as follows.

Method A   Two large panniers suspended from a carrier above the rear wheel. AD is at a height of 70 cm above the ground.

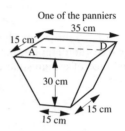

One of the panniers

(a)   Find the distance below AD of the centre of mass of a pannier and hence the height of this centre of mass above ground level.

(b)   Find the volume of luggage in the panniers.

Method B   Two small panniers attached to carriers on the front forks plus a saddlebag. AD is at a height of 50 cm above the ground. The saddlebag is cylindrical, radius 10 cm, length 30 cm and its axis of symmetry is at a height of 80 cm above the ground.

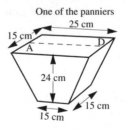
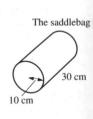

One of the panniers

The saddlebag

(c)   Find
   (i)   the distance below AD of the centre of mass of a pannier
   (ii)   the volume of luggage in the panniers
   (iii)   the volume of luggage in the saddlebag
   (iv)   the height above ground level of the centre of mass of all the luggage.

(d)   Which of the two methods
   (i)   carries the most luggage
   (ii)   has the lower centre of mass?

CHAPTER 7

EQUILIBRIUM OF RIGID BODIES

EQUILIBRIUM OF CONCURRENT OR PARALLEL FORCES

A body that is at rest, or moving with constant linear velocity, is in equilibrium.

In Module E we considered the equilibrium of a body under the action of a set of coplanar forces of two types: concurrent forces and parallel forces

We saw that concurrent forces cannot affect rotation, so the body is in equilibrium provided only that

the resultant force is zero in each of two perpendicular directions.

For parallel forces the resultant force in the direction of those forces must be zero but, in addition,

the resultant moment (turning effect) about *any* axis must also be zero.

Before going further it is helpful to revise these cases and the following exercise provides for this.

EXERCISE 7a

1. The forces shown in the diagram are in equilibrium. Find the values of P and θ.

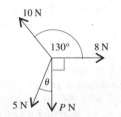

2. A light beam rests in a horizontal position on two supports, one at A and the other at B. It carries loads as shown. Find the force at each support.

3. A particle of weight 8 N is resting in rough
contact with a plane inclined at 60° to the
horizontal. A horizontal force of 5 N is just
able to prevent the particle from slipping down
the plane. Find the coefficient of friction.

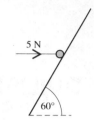

4. The scales in the diagram consist of
a uniform beam ABC of mass
0.5 kg, which is pivoted at B, with
AB = 0.5 m   and   BC = 0.3 m.
A scale plan of mass 0.4 kg is
attached at C and a sliding counter
weight can move on the section AB
in such a way that its centre of
mass can be placed exactly at A. It
is required that masses of up to
10 kg can be weighed when placed
centrally in the scale pan.

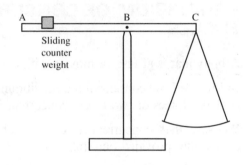

(a) Find the least possible mass for the counter weight.

(b) Find the distance from B of the centre of mass of the counter weight for the
scales to balance when the scale pan is empty.

5. A trolley is made of thin metal bars,
with central cross-section ABC.
Angle ABC $= 90°$, AB $= 0.3$ m
and BC $= 1.2$ m. Its wheels are on
an axle passing through B. A packing
case, of mass 150 kg has a rectangular
cross-section measuring 0.6 m by 0.8 m
and it is packed so that the centre of .
mass is at its centre. The case is
placed on the trolley as shown in the
diagram and a porter supports the
trolley, with BC at an angle θ to the
vertical, by exerting a force P newtons
vertically upwards at C. Assume that
the weight of the trolley is negligible.

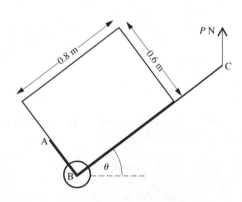

(a) If the porter can place the trolley in a position where it just balances,
without any force P holding it at C, find the value of θ.

(b) If $\theta = 35°$,   find P.

6. A uniform plank of mass 80 kg and length 4 m overhangs a horizontal roof by 1.5 m. A man can walk to within 0.5 m of the overhanging end when a mass of 12 kg is placed on the opposite end. What is the mass of the man and how much bigger a load must be placed at the end of the plank to enable the man to walk right to the overhanging end?

7. A cable car, of mass 800 kg is suspended from the cable on two rods AC and BC, which are inclined at 50° and 35° respectively to the vertical. Find the tensions in the rods AC and BC when the cable car is in equilibrium.

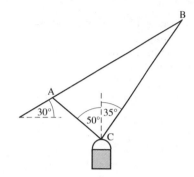

GENERAL CONDITIONS FOR EQUILIBRIUM IN A PLANE

We saw in Module E that when an object is in equilibrium under the action of a set of coplanar forces, there is no change in its linear motion in each of two perpendicular directions and no change in its rotation.

To support these conditions the forces must be such that

> **the resultant force in a specified direction, Ox say, is zero,**
> **the resultant force in the perpendicular direction, Oy, is zero,**
> **the resultant moment about any axis perpendicular to the plane is zero.**

The use in a particular problem of this set of conditions for equilibrium leads to the formation of *not more than three independent equations* relating the forces in the system. If not all the forces need to be considered in a particular problem, it may not be necessary to use all three equations.

It is sometimes convenient to replace one of the first two conditions in the set above, by equating to zero the resultant moment about a second axis, giving

> **the resultant force in a direction parallel to *either* Ox or Oy is zero,**
> **the resultant moment about an axis through A is zero**
> **the resultant moment about another axis through B is zero.**

The use of this alternative set of conditions will be illustrated in one of the following examples which are provided, along with the exercise that follows, as a reminder of the general approach to the equilibrium of a rigid body.

Examples 7b

1. The end A of a uniform rod AB of length $2a$ and weight W is smoothly pivoted to a fixed point on a wall. The end B carries a load of weight $2W$. The rod is held in a horizontal position by a light string joining the midpoint G of the rod to a point C on the wall, vertically above A. The string is inclined at $60°$ to the wall. Find, in terms of W, the tension in the string and the horizontal and vertical components of the force exerted by the pivot on the rod.

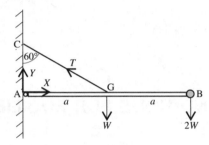

We will represent the components of the force at the pivot by X and Y as shown.

There are three unknown quantities, X, Y and T, so we need three equations and will resolve in two directions and take moments about an axis.

Resolving $\rightarrow$ $X - T \sin 60° = 0$ [1]

Resolving $\uparrow$ $Y + T \cos 60° - W - 2W = 0$ [2]

If we take moments about A, X and Y are not involved.

A$\circlearrowright$ $W \times a + 2W \times 2a - T \times a \sin 30° = 0$ [3]

From [3], $Ta\left(\frac{1}{2}\right) = 5Wa$ $\Rightarrow$ $T = 10W$

Using this value of T in [2] gives

$$Y + 10W\left(\tfrac{1}{2}\right) - 3W = 0 \qquad \Rightarrow \qquad Y = -2W$$

and [1] gives $X - 10W\,\dfrac{\sqrt{3}}{2} = 0$ $\Rightarrow$ $X = 5W\sqrt{3}$

Therefore the tension in the string is $10W$.

The vertical component of the pivot force is $2W$ downwards.

The horizontal component of the pivot force is $5W\sqrt{3}$ acting away from the wall.

If required, the magnitude and direction of the result force at the pivot can be found from the triangle of forces,

2. A ladder of length $2a$ and weight W rests with its foot, A, on rough horizontal ground where the coefficient of friction is $\frac{3}{4}$. The top of the ladder, B, rests against a vertical wall and a painter, of weight $2W$, is standing at the top of the ladder. By modelling the ladder as a uniform rod, the wall as smooth and the painter as a particle, find the angle θ between the ladder and the wall when the ladder is just about to slip. Are there any respects in which the model is inappropriate?

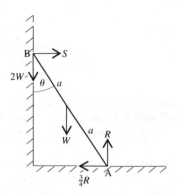

When the ladder is about to slip, the frictional force at A is $\frac{3}{4}R$.

There is no frictional force acting at B.

Resolving horizontally and vertically and taking moments about A give:

$\rightarrow$ $$S - \tfrac{3}{4}R = 0 \qquad\qquad [1]$$

$\uparrow$ $$R - 3W = 0 \qquad\qquad [2]$$

A $\circlearrowleft$ $$W \times a \sin\theta + 2W \times 2a \sin\theta - S \times 2a \cos\theta = 0 \qquad [3]$$

From [1] and [2] $$S = \tfrac{9}{4}W$$

Then [3] gives $$5Wa\sin\theta - \tfrac{9}{2}Wa\cos\theta = 0$$

$\Rightarrow$ $$\tan\theta = 0.9$$

To the nearest degree the ladder is inclined at $42°$ to the wall.

Although it is not unreasonable to model the ladder as a uniform rod, many ladders have their centre of mass below the midpoint and so are not uniform.

It is unrealistic to assume the wall to be smooth; the frictional force that almost certainly acts there will help to prevent the ladder from slipping.

Modelling the painter as a particle at the top of the ladder is not appropriate because

  he probably does not stand on the very top rung, but even if he does he is not at the top of the ladder,

  his centre of mass is unlikely to be vertically over his feet.

3. The diagram shows a drawbridge
   over a moat. It is pivoted along
   one end and chains are attached to
   the outside corners of the other
   end. These chains pass over
   pulleys, immediately above the
   gate, to a winch. Normally the
   drawbridge is raised by operating
   the winch, but unfortunately it has
   broken down and soldiers in the
   castle have been detailed to raise
   the bridge by pulling the chains
   vertically downwards.

Choose a model which will allow you to calculate, in terms of the weight W of the
drawbridge, the least total pull needed when the drawbridge has been raised through
30°. State, with comments on their suitability, any assumptions made.

We will model the drawbridge as a uniform plank (this is rough and ready but
not unreasonable), and treat the chains as light strings (this is not very accurate
as the weight of the chains would cause sagging and also increase the pull
needed). We will also assume that equal pulls are exerted on each side so that
the tensions are equal and the resultant pull acts in the middle. Further, it is
assumed that the height of the pulleys is equal to the length of the drawbridge
i.e. that the drawbridge fills the hole.

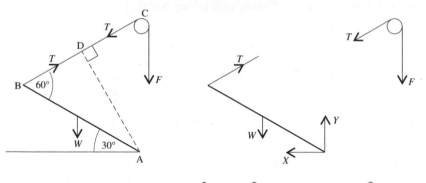

In $\triangle ABC$, $AB = AC = l$ $\Rightarrow$ $A\widehat{B}C = A\widehat{C}B = 60°$ $\Rightarrow$ $B\widehat{A}D = 30°$

The hinge exerts on the drawbridge a force that is unknown both in magnitude
and direction but it can be avoided if we only take moments about A.

A ↺ $W \times \frac{1}{2}l \cos 30° - T \times l \cos 30° = 0$

$\Rightarrow$ $T = \frac{1}{2}W$

Now we will assume that the tension is unchanged by passing over the pulley
(not very reliable as there is certain to be some friction at the pulley which would
increase the tension on the other side).

The least total pull, F, on the chains when the drawbridge has been raised through 30° is given by

$$F = T = \tfrac{1}{3}W$$

Note that the pull is least when the drawbridge is raised steadily and not at an increasing speed. Note also that the pull of $\tfrac{1}{2}W$ is valid only when the drawbridge has been raised through 30°.

4. A wooden plank AB of weight 60 kg and length 4 m rests with A on rough horizontal ground where the coefficient of friction is $\tfrac{1}{2}$. The plank rests in rough contact with the top C of a rail of height 1.5 m, and is just about to slip. Given that AC = 3 m and taking g as 10, find

(a) the normal contact forces at A and at C

(b) the coefficient of friction at C.

Have any assumptions been made?

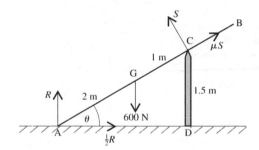

In $\triangle$ACD, $\sin \theta = \dfrac{CD}{AC} = \dfrac{1.5}{3} = \dfrac{1}{2}$

$\therefore \quad \theta = 30°$

As the plank is about to slip, friction is limiting at both A and C.

R and S can be found separately by taking moments about A and C; then we resolve only once. In this problem answers will be given to 2 sf, so in intermediate working we will use 3 sf.

(a) A$\circlearrowleft$ $S \times 3 - 600 \times 2 \cos 30° = 0$

  $\therefore$ $S = 400 \cos 30°$ $\Rightarrow$ $S = 346$

  C$\circlearrowleft$ $600 \times 1 \cos 30° - R \times 3 \cos 30° + \tfrac{1}{2}R \times 3 \sin 30° = 0$

  $\therefore$ $R(\cos 30° - \tfrac{1}{2}\sin 30°) = 200 \cos 30°$ $\Rightarrow$ $R = 281$

  To 2 sf the normal reaction at A is 350 N and that at C is 280 N

(b) Resolving $\leftarrow$ $S \sin 30° - \mu S \cos 30° - \tfrac{1}{2}R = 0$

  $\Rightarrow$ $\mu S \cos 30° = S \sin 30° - \tfrac{1}{2}R$

  $\therefore$ $\mu = \dfrac{173 - 140}{346 \times 0.866} = 0.110 \quad (3 \text{ sf})$

The coefficient of friction at C is 0.11   (2 sf)

It has been assumed that: the plank is uniform and straight; the top of the rail is small enough to be treated as a point.

EXERCISE 7b

In this exercise take all rods, ladders, etc. to be uniform and in limiting equilibrium unless stated otherwise.

1. A ladder, of weight 250 N and length 4 m, rests with one end A on rough horizontal ground and the other end B against a smooth vertical wall. Find

   (a) the normal reaction at B

   (b) the normal reaction at A

   (c) the frictional force at A

   (d) the coefficient of friction with the ground.

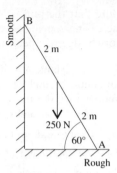

2. A ladder, of weight 200 N and length 5 m, has its foot on rough ground and the top of the ladder rests against a rough wall. The coefficient of friction at both surfaces is $\frac{1}{3}$. If the angle between the ladder and the ground is θ, find the value of θ when the ladder is on the point of slipping.

   (Remember that the ladder will not slip until friction is limiting at both points of contact.)

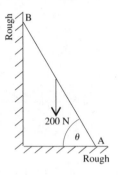

3. The diagram shows a central cross-section AB of a window which has a mass of 4 kg. The window is held at an angle of 30° to the vertical by a light rod BC, which is attached to the window frame at C. AB = 0.3 m and angle ABC = 90°. Find

   (a) the tension in BC

   (b) the magnitude and direction of the reaction at the hinge.

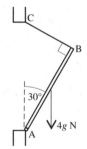

4. A uniform rod XY whose mid point is M, is in equilibrium in a vertical plane as shown in the diagram.

   The rod rests on a rough peg at Z and a force F acts at X as shown.
   If YZ = ZM and $\tan \alpha = \frac{4}{3}$, find the coefficient of friction at Z and the force F.

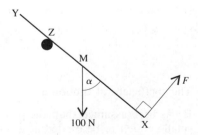

5. In the diagram AB is a table top, of weight
   49 N, hinged to a vertical wall at A. The
   table top is supported by a light rod CD,
   which is hinged to the wall at D.
   AC = CB = 0.3 m and AD = 0.4 m.
   A boy leans on the table at B exerting a force
   of 8 N vertically downwards. Find

   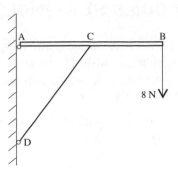

   (a) the thrust exerted on the table top by
       the rod CD
   (b) the magnitude and direction of the
       reaction at the hinge A.

6. A uniform rod XY, whose weight is W,
   is in equilibrium in a vertical plane. The
   midpoint of the rod is at M. The rod is
   supported on a plane inclined at 30° to
   the horizontal, by a string attached to
   the end X and held vertically. The rod,
   whose weight is W, is inclined to the
   plane at 30° as shown in the diagram.
   Find

   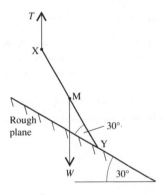

   (a) the tension in the string in terms
       of W
   (b) the coefficient of friction between
       the rod and the plane.

7. The diagram shows a uniform rectangular shelf ABCD, of mass 5 kg. It is hinged
   to a wall along the edge AD and supported by light rods BK and CL, which are
   attached to the wall at K and L. AB = 0.4 m and AK = DL = 0.3 m.
   Find

   (a) the tension in each of the rods BK and CL
   (b) the magnitude and direction of the reaction at the hinge.

   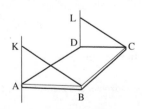
   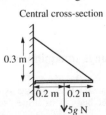

8. A light ladder, of length 4 m stands on rough horizontal ground, with coefficient
   of friction 0.25, and its upper end rests against a smooth vertical wall. The
   ladder is inclined at 65° to the horizontal. The end rungs are each 0.3 m from an
   end of the ladder. A man, of mass 80 kg, stands on the top rung and another
   man, of mass m kg, stands on the bottom rung. Find the least value of m which
   will prevent the ladder from slipping.

BODIES SUSPENDED IN EQUILIBRIUM

When a body whose centre of mass is at G is suspended from a point P, it rests in equilibrium with G vertically below P. In Module E we considered the suspension of a lamina and now we extend the method used there to deal with rigid bodies.

Clearly the equilibrium aspect cannot be dealt with until the location of G is known so, unless the body is one for which G can be quoted, we must first calculate the position of G.

Examples 7c

1. **A solid right circular cone, of height 4a and base radius a, is suspended freely from a point P on the circumference of the base. Find the angle α between PO and the vertical, where O is the centre of the base.**

The centre of mass of a solid cone is known to be on the line of symmetry, $\frac{3}{4}$ of the way from the vertex to the base.

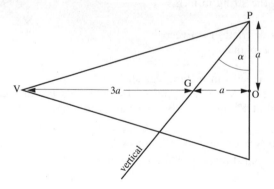

When the cone hangs in equilibrium, PG is vertical.

It is often easier to draw the body with its axis of symmetry horizontal and *mark* the line GP as vertical rather than to draw the body in its suspended position.

G is distant 3a from the vertex and a from O.

Therefore $\tan \alpha = \dfrac{GO}{OP} = \dfrac{a}{a} = 1$

The angle between OP and the vertical is 45°.

2. A sculpture is in the form of a uniform solid cylinder of radius 2 cm and mass $4M$, with a small lead bead of mass M let in at a point A on the rim of the base. If the sculpture is suspended from the point B, directly above A on the upper rim, AB is inclined to the vertical at an angle α whose tangent is $\frac{1}{3}$. Find the height, AB, of the sculpture.

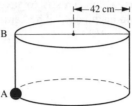

Let the height of the cylinder be $2h$ centimetres.

First we need to find the centre of mass of the sculpture.

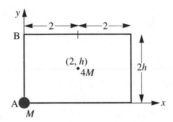

		Coords of C of M			
Portion	Mass	x	y	Σmx	Σmy
Cylinder	$4M$	2	h	$8M$	$4Mh$
Lead bead	M	0	0	0	0
Sculpture	$5M$	$\bar{x}$	$\bar{y}$	$5M\bar{x}$	$5M\bar{y}$

Using $\Sigma mx = \bar{x}\,\Sigma m$ gives $8M = 5M\bar{x}$ $\Rightarrow$ $\bar{x} = \frac{8}{5}$

Using $\Sigma my = \bar{y}\,\Sigma m$ gives $4Mh = 5M$ $\Rightarrow$ $\bar{y} = \dfrac{4h}{5}$

$\mathrm{B\widehat{G}}$ is vertical therefore $\mathrm{A\widehat{B}G}$ is α.

$\therefore$ $\tan \alpha = \dfrac{\bar{x}}{2h - \bar{y}} = \dfrac{\frac{8}{5}}{\frac{6h}{5}} = \dfrac{4}{3h}$

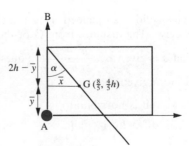

But $\tan \alpha = \frac{1}{3}$

$\therefore$ $h = 4$

The height of the sculpture is 8 cm.

EXERCISE 7c

Most of the questions in this exercise refer to objects whose centres of mass were found in Exercise 6d and, for interest, the question reference is given. So that you can concentrate on *how the objects hang when suspended*, the position of the centre of mass of each body is given. This is to save you having to begin by calculating the position of G in each case. The centre of mass of each solid is on the axis of symmetry.

1. The centre of mass of this solid (question 2a) is $\frac{5}{3}$ cm from O. If the solid is suspended from A, find the angle between AB and the vertical.

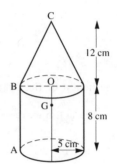

2. The centre of mass of the frustum of a cone shown in the diagram is 4.7 cm above the base. Find the angle between the axis of symmetry and the vertical when the frustum is suspended freely from the point P as indicated.

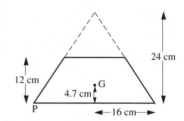

3. The solid shown in the diagram is freely suspended from C, the midpoint of AB. Find the angle between AB and the vertical given that the height above the base of the centre of mass is $\frac{83}{88}a$ (question 10).

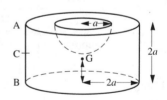

4. This solid is suspended from B and hangs freely. The distance from O of the centre of mass is given by $\dfrac{h^2 - 24}{2(h+2)}$ (question 4).

   Find h given that

   (a) AB is horizontal

   (b) BC is horizontal.

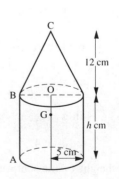

BODIES RESTING IN EQUILIBRIUM ON A HORIZONTAL PLANE

Consider a solid body, with constant cross-section, resting with one face on a horizontal plane. Depending on the shape of the cross-section, the body may have a tendency to topple over. The solid whose cross-section is shown in the following diagrams, for example, could topple over about B to rest on the edge through C, but whether or not it does fall over depends on the relative positions of B and G.

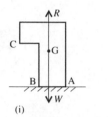 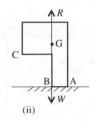 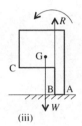

(i)   (ii)   (iii)

In diagram (i) the weight W of the body acts through a point within the face through AB and the normal contact force R also acts through the same point, so the body can remain in equilibrium.

In the extreme position, diagram (ii), W and R act through B itself, keeping the body *just* in equilibrium.

If we consider moments about an axis through B, both W and R have zero moment. In this case we can see that the weight of the shaded part exerts an overturning moment, while the weight of the unshaded portion exerts a restoring moment about B and these must be equal.

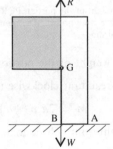

Now in diagram (iii) on the other hand, if AB *could* remain in contact with the plane, R would act through some point on AB and therefore the moments of W and R about B would both be in the overturning sense. Therefore the body will topple over (there will then be two normal reactions, one through B and another through C).

So the question of whether the body will rest in the given position can be answered either by checking the resultant moment about B or by checking that the overall weight passes through a point within the base of contact (this may involve having to find the centre of gravity of the object).

Examples 7d

1. A uniform prism whose cross-section is the trapezium shown in the diagram, is placed with the side PQ on a horizontal plane. Find the range of values of k for which the prism will rest in equilibrium in this position.

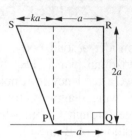

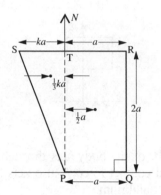

Consider the extreme position when the normal reaction passes through P.

The weight of the portion PQRT exerts a clockwise moment about an axis through P, i.e. in the sense to preserve equilibrium. The moment of the weight of the portion PTS is anticlockwise and tends to cause overturning.

The weight of the portion PTS is $ka^2\rho$ and that of PQRT is $2a^2\rho$.

The resultant clockwise moment, M, about an axis through P is given by

$$M = 2a^2\rho\left(\tfrac{1}{2}a\right) - ka^2\rho\left(\tfrac{1}{3}ka\right) = \tfrac{1}{3}(3 - k^2)a^2\rho$$

Equilibrium will be maintained as long as M does not become negative,

i.e. for equilibrium $M \geqslant 0$

$\Rightarrow$ $3 - k^2 \geqslant 0$

In this problem k cannot be negative so, for equilibrium,

$$0 \leqslant k \leqslant \sqrt{3}$$

2. A man has invented a game that is played with a wood in the shape of a frustum of a cone. The wood is bowled with its curved surface in contact with the ground so that it describes a circle. The man is cutting his wood from the solid cone with base radius 12 cm and height 16 cm, shown in the diagram.

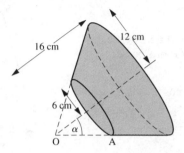

He plans to make a cut 8 cm from the vertex so as to remove the top half of the cone and wants to check that the wood will not fall over on to its smaller face when it is placed on horizontal ground.

(a) Find the length of OA.

(b) Write down, as a fraction, the value of cos α.

(c) Find the length of OG where G is the centre of mass of the wood.

(d) Given that the weight of the wood passes through the point H on the plane, find the length of OH.

(e) Determine whether the wood will topple on to its smaller face.

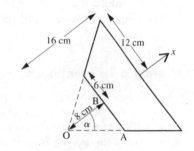

(a) OB $= \frac{1}{2}$ (height of whole cone)

   $\Rightarrow$ AB $= \frac{1}{2}$ (radius of whole cone) $= 6$ cm

   Using Pythagoras' theorem in $\triangle$OAB gives
   OA $= 10$ cm

(b) cos α = OB/OA $= \frac{8}{10} = \frac{4}{5}$

(c) The linear ratio of the similar cones is $8:16$, i.e. $1:2$, so the ratio of their volumes, and therefore their masses, is $1:8$. Let their masses be M and $8M$.

Portion	Mass	x coord of centre of mass	Σmx
+ Large cone	$8M$	$\frac{3}{4} \times 16$	$96M$
− Small cone	M	$\frac{3}{4} \times 8$	$6M$
Wood	$7M$	$\bar{x}$	$7M\bar{x}$

Using $\Sigma mx = \bar{x}\,\Sigma m$ gives $96M - 6M = 7M\bar{x}$

   $\Rightarrow$ $\bar{x} = \frac{90}{7} = 12.9\ldots$

i.e.   OG $= 13$ cm   (2 sf)

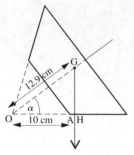

(d) The weight of the wood acts vertically downwards through G and passes through the point H,

$$\therefore \qquad OH = OG \cos \alpha$$

i.e. $\qquad OH = \frac{90}{7} \times \frac{4}{5} = 10.3 \, \text{cm}$

(e) $OH = 10.3 \, \text{cm}$ and $OA = 10 \, \text{cm}$, i.e. $OA < OH$

so H is a point inside the line of contact between the wood and the plane.

Therefore the wood will not topple over when placed on the plane.

The examples we have looked at so far have had either a plane or a line in contact with the horizontal plane. If an object with a spherical surface rests in contact with a plane, there is only a point of contact. The normal contact force therefore must pass through this point so the situation is similar to that of suspension, i.e. the point of contact must be vertically below the centre of gravity. Additionally we know that, because the horizontal plane is tangential to the circular cross-section, the normal reaction must pass through the centre of the circle.

Examples 7d (continued)

3. **The diagram shows the central cross-section of a casting in the shape of a hemisphere of radius 8 cm, with a hemispherical depression, of radius 4 cm. When the casting rests in equilibrium with its curved surface touching a horizontal surface, find the inclination to the horizontal of the line of symmetry (OAB) of the plane face.**

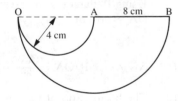

First we will find the centre of mass of the casting, knowing that it lies in the section of symmetry shown in the diagram.

The two hemispheres are similar with a linear ratio $2:1$, so their masses are in the ratio $8:1$.

Portion	Mass	Coords of G		mx	my
		x	y		
+ Large hemisphere	$8M$	8	$\frac{3}{8} \times 8$	$64M$	$24M$
− Small hemisphere	M	4	$\frac{3}{8} \times 4$	$4M$	$\frac{3}{2}M$
Casting	$7M$	$\bar{x}$	$\bar{y}$	$7M\bar{x}$	$7M\bar{y}$

Using $\Sigma mx = \bar{x}\,\Sigma m$ gives $64M - 4M = 7M\bar{x}$

Using $\Sigma my = \bar{y}\,\Sigma m$ gives $24M - \frac{3}{2}M = 7M\bar{y}$

$\Rightarrow$ $\bar{x} = \frac{60}{7}$ and $\bar{y} = \frac{45}{14}$

When the casting rests in equilibrium on a horizontal plane, touching at P, GP is vertical. Also as the plane is tangential to the large hemisphere, GP passes through its centre A.

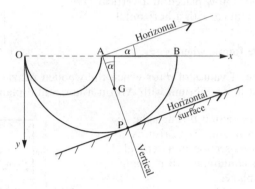

Therefore $\tan\alpha = \dfrac{\bar{x} - 8}{\bar{y}} = \dfrac{4}{7} \div \dfrac{45}{14} = \dfrac{8}{45}$

$\therefore$ $\alpha = 10°$ to the nearest degree.

EXERCISE 7d

1. The diagram shows the cross-section through the centre of mass of a uniform prism. The prism has been placed with the face containing AB in contact with a horizontal plane.

   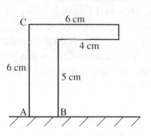

   (a) Find the distance of the centre of mass of the prism from AC.

   (b) Can the prism rest in equilibrium without toppling over?

2. ABCDE is a cross-section through the centre of mass of a uniform prism, which is placed with the face containing AB in contact with a horizontal plane.

   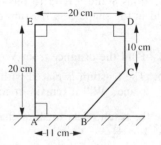

   (a) Find the distance of the centre of mass of the prism from AE.

   (b) Determine whether the prism will rest in equilibrium or will topple.

3. The prism described in question 2 is altered by reducing AB to 8 cm without changing the other dimensions. Find out whether it can now rest in equilibrium with AB on the horizontal plane.

4. The diagram shows a piece of a wooden puzzle.

   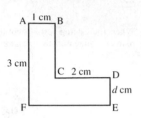

   (a) Find, in terms of d, the distance from AF of its centre of mass.

   The piece of puzzle is now placed in a vertical plane with AB in contact with a horizontal plane.

   (b) Will the piece topple when $d = 2$?

5. Calculate the range of values of d for which the wooden puzzle piece in question 4 can rest in equilibrium with AB on a horizontal plane.

6. ABCDEF is a cross-section through the centre of mass of a prism. The prism is placed with this cross-section in a vertical plane and the face containing AB in contact with a horizontal plane.

   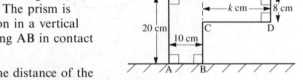

   (a) Find, in terms of k, the distance of the centre of mass of the prism from AF.

   (b) Determine whether the prism can stay in equilibrium in this position without toppling if (i) $k = 5$ (ii) $k = 20$

   (c) Find the range of values of k for which equilibrium is possible.

7. The solid in the diagram consists of a cylinder and a cone attached to each other at their common plane faces. The centre of mass is at the point G, where $OG = \frac{15}{4}$ cm. Determine whether the solid will remain in equilibrium when it is placed with BC in contact with a horizontal plane, or will topple over on to AB.

   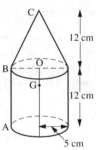

Questions 8 and 9 refer to this diagram of a uniform solid in the shape of a frustum of a cone.

8. (a) Find the distance from V of its centre of mass.

   (b) The frustum is placed with its curved surface in contact with a horizontal plane. Will it remain in equilibrium in this position if $\alpha = 35°$.

*9. Use the result of 9a to show that the frustum is on the point of toppling when $\cos^2 \alpha = \frac{13}{30}$.

EQUILIBRIUM ON AN INCLINED PLANE

One of the properties necessary for the equilibrium of a body resting on a horizontal plane, applies also to a body in equilibrium on an inclined plane, namely:

If there is a plane or a line of contact, the weight, which acts vertically downwards through G, must pass through a point within that region of contact; the normal reaction acts through that point.

On an inclined plane however, the normal reaction, which is perpendicular to the inclined plane, is not vertical and is therefore not collinear with the weight. The normal reaction has a horizontal component but the weight of the body does not. Therefore equilibrium is not possible unless another force with a horizontal component acts on the body. So, unless an extra supporting force is applied to the body, there must be friction between the body and the plane.

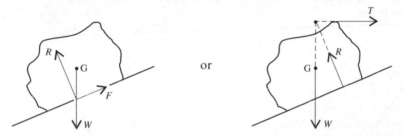

or

Note that, in either case, the three forces are concurrent at a point on the line of action of the weight.

Note also that if a body has a spherical surface, it has contact with the plane at only one point so the weight must pass through that point.

When friction maintains equilibrium, the *resultant* contact force S (i.e. the resultant of the friction and the normal reaction) balances the weight; it is therefore vertical and must pass through G.

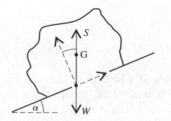
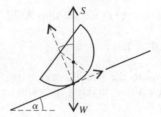

In general, then, if a body is placed on an inclined plane without any supporting force, its equilibrium depends not only upon the position of G relative to the contact region, but also on whether slipping can occur.

When slipping is about to occur, it can be seen from the diagram that the angle of friction (between R and S) is equal to α.

Examples 7e

1. A uniform solid cylinder, with radius a and height $3a$, is resting in equilibrium with one end on a rough plane inclined at an angle α to the horizontal. The inclination of the plane is gradually increased until the cylinder is just on the point of toppling over.

   (a)   Find the greatest possible value of α

   (b)   Find the least value of the coefficient of friction between the plane and the cylinder.

(a)

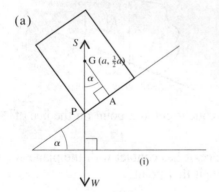
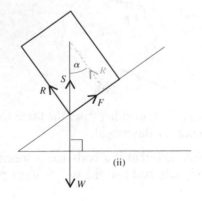

The cylinder is in equilibrium under the action of two forces, its weight and the resultant contact force, therefore these two forces are collinear. The cylinder is on the point of toppling over so the contact forces act through the lowest point P of its plane face.

From diagram (i) we see that $A\widehat{G}P = \alpha$

In $\triangle AGP$, $\tan \alpha = a \div \frac{3}{2}a = \frac{2}{3}$ $\Rightarrow$ $\alpha = 33.6 \ldots ^\circ$

Therefore, to the nearest degree, the largest angle at which the plane can be elevated without making the cylinder topple over is $33°$ ($34°$ would cause toppling).

(b) In diagram (ii) we see that $F = R \tan \alpha$

But $F \leqslant \mu R$ $\Rightarrow$ $\mu \geqslant \tan \alpha$

Therefore just to prevent slipping at the maximum elevation, $\mu = \frac{2}{3}$

2. **At a warehouse, packages of goods are delivered from the first floor store to the ground floor for despatch, by placing them on a moving ramp. All the goods in one section are packaged in boxes that are cuboids, all with a base 1.6 m square but of varying heights. The ramp, which is rough enough for there to be no possibility of a box slipping, slopes down at 40° to the horizontal. By modelling the boxes as uniform cuboids, find the maximum permitted height of a box for safe delivery.**

   State, with reasons, whether you think that the model is appropriate and suggest ways, if any, in which it might be improved.

   The boxes cannot slip, so the only restriction necessary is to ensure that no box topples over.

   The weight acts through the centre of mass G and, for equilibrium, the resultant contact force must act in the same line.

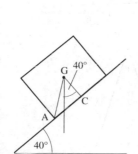 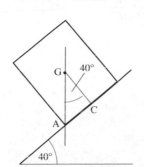 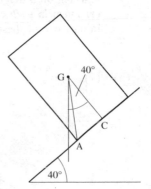

For equilibrium, the angle AGC must be greater than or equal to 40°

i.e. $\tan \text{AGC} \geqslant \tan 40°$

Taking the height of a box as $2h$ gives $\tan \text{AGC} = \dfrac{0.8}{h}$ $\Rightarrow$ $\dfrac{0.8}{h} \geqslant \tan 40°$

$\Rightarrow$ $h \leqslant 0.95$ (to 2 sf, rounded down for safety)

Therefore the maximum permitted height should be 1.9 m.

The suitability of the model depends on whether the contents of a box are always such that the centre of mass is halfway up. Also the conveyor belt may not run smoothly and jerks could make toppling more likely even at 'safe' heights.

Taking a centre of mass at, say, 60% of the height of the box might give more reliable results.

3. The diagram shows the cross-section of a uniform solid
   brass ornament in the shape of a hemisphere with a cone
   attached to its plane face. The radius of both hemisphere
   and cone is $2a$.

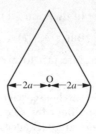

(a) Sketch the ornament if the height of the cone is (i) a (ii) $6a$ and mark on
    each sketch an estimate of the position of the centre of mass, G, of the
    ornament.

(b) It is intended to display the ornament with its
    hemispherical surface resting on a velvet covered
    inclined shelf as shown.

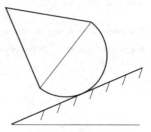

   Use your estimated positions for G to explain, by
   marking on a diagram the forces that act on the
   ornament in each case, that equilibrium is possible
   in only one of these cases.

(a) (i)   (ii)

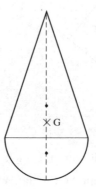

Mass of cone $<$ mass of hemisphere   Mass of cone $>$ mass of hemisphere
G lies within the hemisphere less   G lies within the cone,
than $\frac{3}{4}a$ from the centre.   less than $\frac{3}{2}a$ from its base.

(b) Assume that contact with the velvet is rough enough to prevent slipping.

   The normal reaction passes through the centre of the common base and
   friction acts up the plane. So the resultant contact force, S, is inclined
   'uphill' from P.

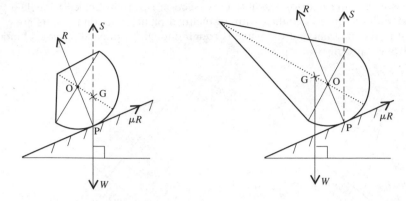

In case (i) G is right of, and below, the centre O so the resultant contact force *can* pass through G producing equilibrium.

In case (ii) G is left of, and above, the centre O so the resultant contact force *cannot* pass through G and equilibrium is impossible.

EXERCISE 7e

1. A cube with sides of length 20 cm is placed with its plane face in contact with an inclined plane, which is rough enough to prevent sliding. Determine whether the cube will rest in equilibrium, or topple.

(a)   (b)

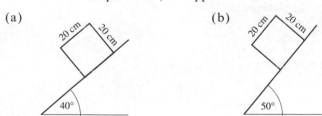

In questions 2 to 4 a uniform solid cone, with dimensions shown in the diagram, rests with its plane face in contact with a plane that is rough enough to prevent sliding. Determine whether the cone will topple.

2.   3.   4.

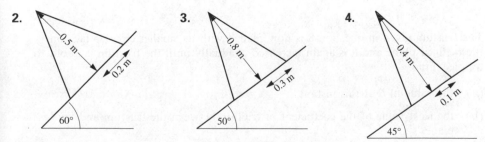

In questions 5 and 6 a prism, whose cross-section is a right-angled triangle is placed with one face in contact with an inclined plane. The dimensions are shown in the diagram and the plane is rough enough to prevent sliding. Find whether the prism will topple.

5.

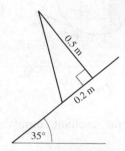

6.

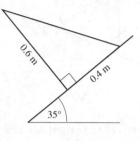

7. A uniform solid cylinder of base radius $10\,\text{cm}$ and height $h\,\text{cm}$ is just on the point of toppling when it is placed with one end on a rough plane inclined at an angle of $30°$ to the horizontal. Find the value of h.

8. The larger circular face of the frustum of a cone shown in the diagram is placed on a rough plane inclined at θ to the horizontal. The inclination of the plane is steadily increased until, when $\theta = 60°$, the frustum is on the point of toppling. Find the radius of each of the plane faces.

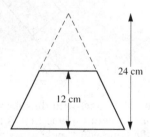

9. The frustum given in question 8 is now placed with its smaller circular face on the inclined plane and θ is again increased gradually until the frustum is about to topple. Find

(a)   the value of θ at this instant

(b)   the least value of the coefficient of friction between the frustum and the plane.

*10. The diagram shows a solid formed by joining the plane faces of a cone, of radius a and height a, to a hemisphere of radius a.

The centre of mass of the compound body is distant $\frac{1}{6}a$ from O. The body rests in equilibrium with its axis of symmetry horizontal on a rough inclined plane as shown.

Find the value of θ.

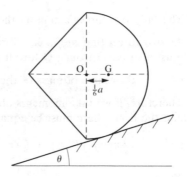

FURTHER PROBLEMS

In this section we look at some problems which, while using the same mechanical principles as have been applied so far, involve more analysis and are a little harder. Anyone who enjoys dipping a little deeper into a subject will find them interesting and rewarding.

We give a few examples to illustrate some of the possibilities, and the exercise that follows includes more variations.

Examples 7f

1. A uniform solid consists of a hemisphere of radius r and a right circular cone of base radius r fixed together so that their plane faces coincide. If the solid can rest in equilibrium with *any* point of the curved surface of the hemisphere in contact with a horizontal plane, find the height of the cone in terms of r.

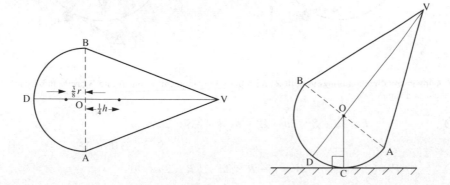

The plane is tangential to the hemisphere so the normal reaction acting at the point of contact, C, passes through O no matter where C is on the surface of the hemisphere.

The only two forces acting on the body are its weight and the normal reaction.

As two forces that are in equilibrium must be collinear, the weight of the solid must also pass through O for all positions of C,

i.e.   the centre of mass of the solid *must be at O*.

Therefore, if we take moments about an axis through O, the moments of the cone and the hemisphere must be equal and opposite,

i.e. $\left(\frac{1}{3}\pi r^2 h\rho\right)\left(\frac{1}{4}h\right) = \left(\frac{2}{3}\pi r^2\rho\right)\left(\frac{3}{8}r\right)$

$\Rightarrow$ $h^2 = 3r^2$

$\therefore$ $h = r\sqrt{3}$

2. A circular disc with centre O, radius $2a$ and weight W, rests in a vertical plane on two rough pegs A and B. OA and OB are inclined to the vertical at $60°$ and $30°$ respectively. Given that the coefficient of friction is $\frac{1}{2}$ at each peg, find the greatest force that can be applied tangentially at the highest point of the disc without causing rotation. Give the answer in surd form and in terms of W.

When rotation is about to take place the disc is about to slip at both pegs, i.e. friction is limiting at both pegs.

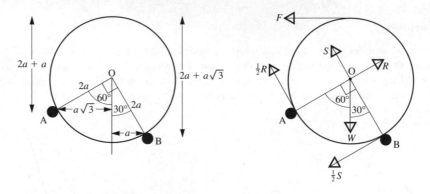

Resolving in any direction brings in both R and S but taking moments about A and B brings in only one of them at a time.

A$\circlearrowleft$ $F \times 3a + S \times 2a - W \times a\sqrt{3} - \frac{1}{2}S \times 2a = 0$

$\Rightarrow$ $S = \sqrt{3}W - 3F$ [1]

B$\circlearrowleft$ $F \times (2+\sqrt{3})a + W \times a - R \times 2a - \frac{1}{2}R \times 2a = 0$

$\Rightarrow$ $3R = W + F(2+\sqrt{3})$ [2]

O$\circlearrowleft$ $F \times 2a - \frac{1}{2}S \times 2a - \frac{1}{2}R \times 2a = 0$

$\Rightarrow$ $2F = S + R$ [3]

Using [1] and [2] in [3] × 3 gives

$$6F = 3\sqrt{3}W - 9F + W + 2F + \sqrt{3}F$$

$$\Rightarrow \quad F(13 - \sqrt{3}) = W(3\sqrt{3} + 1)$$

The greatest force is $\dfrac{(3\sqrt{3} + 1)W}{(13 - \sqrt{3})}$

Note that taking moments about three different axes gives three independent facts *provided that the axes are not collinear.* In this problem A, O and B are not collinear.

3. **The cross-section of a uniform prism is a trapezium ABCE. This trapezium is formed from a square lamina, ABCD, with the portion ADE removed. The side of the square is 2 m and the length of ED is 1.5 m.**

   (a) **If the prism is placed with ABCE in a vertical plane and AE on a rough horizontal plane, show that it will topple about the edge through E.**

   (b) **Find the least force that must be applied at C to prevent toppling.**

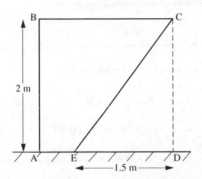

The prism will topple if the line of action of the weight does not pass through a point within AE; to check this we need only the distance from AB of the centre of mass of the cross-section.

Portion	Mass (m)	Distance from AB of C of M (x)	mx
ABCE	4ρ	1	4ρ
−DCE	1.5ρ ·	1.5	2.25ρ
ABCE	2.5ρ	$\bar{x}$	$2.5\rho\bar{x}$

Using $\Sigma mx = \bar{x}\,\Sigma m$ gives

$$4\rho - 2.25\rho = 2.5\rho\bar{x} \quad \Rightarrow \quad \bar{x} = \frac{1.75}{2.5} = 0.7$$

(a) AE = 0.5 which is less than $\bar{x}$, therefore the line of action of the weight of the prism does not intersect AE and the prism will topple about the edge through E.

(b) Toppling about E is caused by the moment of the weight about E. In order just to prevent toppling, the moment about E of the force F newtons applied at C must counterbalance the moment of the weight. The moment of the force is given by $F \times$ perpendicular distance from E. Therefore F will be least when this distance is greatest and this is when the line of action of F is perpendicular to EC.

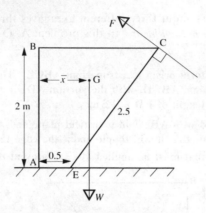

Taking W newtons as the weight of the prism,

E$\circlearrowright$ $W \times (\bar{x} - 0.5) - F \times EC = 0$

$\Rightarrow$ $2.5F = W(0.7 - 0.5) = 0.2W$

$\Rightarrow$ $F = \frac{2}{25}W$

The least force required is $0.08W$ newtons.

4. The diagram shows the central cross-section of a uniform cube which is placed on a rough plane inclined at α to the horizontal where $\tan \alpha = \frac{1}{2}$. A horizontal force P of gradually increasing magnitude is applied at D as shown. If $\mu = \frac{2}{3}$, show that the cube will begin to turn about the edge through B before it begins to slide up the plane.

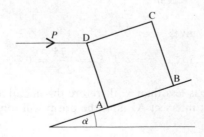

Consider separately the values of P for which sliding or overturning would begin.

Suppose that sliding up is about to begin when $P = P_1$.

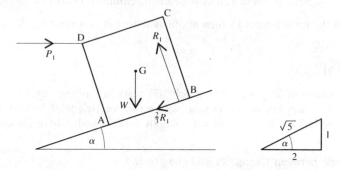

$\uparrow$ $R_1 \cos \alpha - \dfrac{2}{3} R_1 \sin \alpha - W = 0$ $\Rightarrow$ $R_1 \left(\dfrac{2}{\sqrt{5}} - \dfrac{2}{3} \times \dfrac{1}{\sqrt{5}} \right) = W$ [1]

$\rightarrow$ $P_1 - R \sin \alpha - \dfrac{2}{3} R \cos \alpha = 0$ $\Rightarrow$ $R_1 \left(\dfrac{1}{\sqrt{5}} + \dfrac{2}{3} \times \dfrac{2}{\sqrt{5}} \right) = P_1$ [2]

From [1] $R_1 = \dfrac{3\sqrt{5}W}{4}$

From [2] $P_1 = \dfrac{7R}{3\sqrt{5}}$ $\Rightarrow$ $P_1 = \dfrac{7}{4} W$

Now suppose instead that $P = P_2$ and the cube is just about to overturn about B.

To make taking moments easier, both W and P_2 are replaced by their components parallel and perpendicular to the plane as shown.

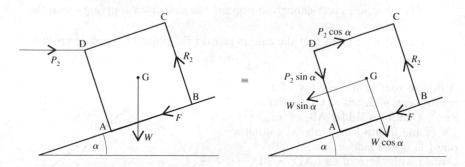

Friction at B is not necessarily limiting but we can avoid the unknown frictional force by taking moments about B.

B$\circlearrowright$ $P_2 \cos \alpha \times 2a - P_2 \sin \alpha \times 2a - W \cos \alpha \times a - W \sin \alpha \times a = 0$ [3]

$\therefore$ $P_2 \dfrac{(4-2)}{\sqrt{5}} = W \dfrac{(2+1)}{\sqrt{5}}$ $\Rightarrow$ $P_2 = \dfrac{3W}{2}$

The values of P_1 and P_2 are different so the cube cannot begin to slide and topple simultaneously.

As soon as the lower of these values is reached, equilibrium will be broken.

$P_2 < P_1$ so the cube begins to turn about B before it can slide up the plane.

EXERCISE 7f

1. A uniform ladder, of length $2a$ and weight W, rests in limiting equilibrium with its foot, A, on rough horizontal ground where the coefficient of friction is $\frac{1}{3}$. The top of the ladder, B, rests against a rough vertical wall where the coefficient of friction is $\frac{1}{2}$.

   Find the angle between the ladder and the ground.

2. A packing case, of mass 40 kg, is in the shape of a cuboid measuring 2 m by 1 m by 1.5 m. It can be assumed that it is packed to a uniform density. A man places it on a moving ramp which is inclined at an angle θ to the horizontal and is rough enough to prevent slipping. The 2 m by 1 m face is in contact with the ramp so that, in cross-section, the case appears as in the diagram.

   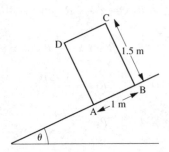

   (a) Show that it cannot rest in equilibrium in this position without toppling if $\tan\theta > \frac{2}{3}$.

   (b) If $\theta = 35°$ and the man applies a force P newtons at D in the direction DC, find P if
   - (i) P is just great enough to prevent the case from toppling down the ramp
   - (ii) P is so great that the case is just on the point of toppling up the ramp.

3. A domed roof is in the shape of a hemisphere, with centre C and radius 8 m. A uniform ladder AB, of weight 200 N and length 8 m, is placed so that a point D of the ladder is in smooth contact with the roof and AD = 6 m. The foot of the ladder, A, is resting on rough horizontal ground level with the base of the hemisphere. The ladder is on the point of sliding.

   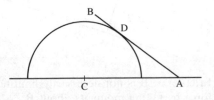

   Find the coefficient of friction between the ground and the ladder.

4. When the uniform solid shown in the diagram is
placed with a point of its hemispherical surface in
contact with a horizontal plane as shown in the
diagram, one of the following things will happen:

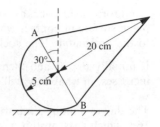

   (i) it will remain in equilibrium in this position
   (ii) it will rotate until AB is horizontal
   (iii) it will rotate until it topples about B

Find which of these three occurs.

5. Carry out the same investigation as that required in
question 4, for another solid of the same type but
with different dimensions as shown.

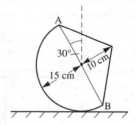

6. The dimensions of a third solid of the same type are
shown in the diagram.

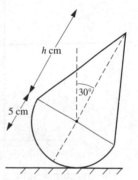

   (a) Find the value of h so that the solid will
remain in equilibrium in the position shown.

   (b) If h has the value found in part (a), describe
what will happen if the solid is placed on the
plane with its axis of symmetry at 80° to the
vertical and then released.

7. A uniform rod AB, of weight W and length $4a$, rests in
limiting equilibrium at an angle θ to the horizontal in
rough contact with two pegs, as shown in the diagram.
One peg is at B and the other at a point C on the
rod where AC $= 3a$. Find, in terms of θ, the
coefficient of friction, which is the same at both pegs.

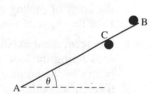

8. A uniform solid cone, of weight W, base radius a and height $4a$, is placed with its
plane face in contact with a rough horizontal plane. The coefficient of friction
between the cone and the plane is $\frac{3}{4}$. A horizontal force P is applied to the cone
half-way up its height.

   (a) Assuming that it does not topple first, find P when the cone is just on the
point of sliding.

   (b) Assuming that it does not slide, find P when the cone is just on the point of
toppling.

   (c) If initially $P = 0$ and then the value of P is gradually increased, in what
way will equilibrium be broken?

9. A uniform solid cone, of weight W, base radius a and height $2a$, is placed with its plane surface in contact with a rough plane which can be inclined at an angle θ to the horizontal. The coefficient of friction between the cone and the plane is $\frac{1}{4}$.

If initially $\theta = 0$ and then the plane is gradually tilted so that the value of θ increases, in what way will equilibrium be broken?

10. The diagram models a tower crane which consists of a gantry ABCD, of length $16a$, which rests on top of a vertical tower. The gantry is of mass $10M$. It has a counter weight centred on end A and a trolley of mass M can move along section CD. Loads are carried suspended from the trolley on a cable.

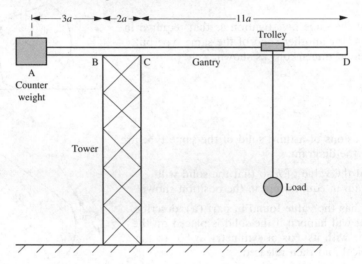

(a) The counter weight is such that, if the trolley were not fitted, it would be on the point of pulling end A of the gantry downwards, turning about point B. Find the mass of the counter weight.

(b) The theoretical maximum load that the crane could lift is determined by considering the load suspended below D which would bring end D of the gantry to the point of turning downwards about point C. Find this theoretical load.

(In practice the trolley cannot reach this position and other safety margins would be built in.)

CONSOLIDATION B

SUMMARY

Motion in a Vertical Circle

A particle travelling round a circular path in a vertical plane has

- an acceleration component towards the centre, of $r\omega^2$ where ω is not usually constant,

- a tangential acceleration component of $r\left(\dfrac{d^2\theta}{dt^2}\right)$

If the particle cannot leave the circular path (e.g. a bead on a circular wire) it can describe either complete circles or an arc of any size. For the particle to perform complete circles there must be a positive velocity at the highest point of the circle. Finding the condition for this to apply requires using only the conservation of mechanical energy (provided that no external work is being done).

If there is nothing physical to prevent the particle from leaving the circle, e.g. if it is rotating at the end of a string which can go slack, the particle may describe complete circles or it may oscillate through an arc that is less than a semicircle, or it may leave the circular path and then travel as a projectile.

The condition for complete circles to be described is found by checking that the force, other than the weight, acting along the radius towards the centre (e.g. the tension in the string) does not become zero before the highest point of the circle is reached, (e.g. that $T \geqslant 0$ at the highest point of the circle).

Elastic Impact

If a collision results in a bounce, the impact is elastic.

Newton's Law of Restitution states that, for two particular colliding particles, the ratio of their relative speed after impact to their relative speed before impact is constant, i.e.

separation speed $= e \times$ approach speed

where e is the coefficient of restitution between the two particles.

In general $0 \leqslant e \leqslant 1$

If $e = 0$ the impact is inelastic and the particles do not bounce.

If $e = 1$ the impact is said to be perfectly elastic and no loss in KE is caused by the collision.

At any impact an impulse acts on each colliding object; the magnitude of the impulse is found by considering the change in momentum of *one* object only.

When both of the colliding objects are free to move, the total momentum in any specified direction is unchanged by the impact.

Centre of Mass of a Rigid Body

The C of M of a uniform solid of revolution lies on the axis of rotation, which is an axis of symmetry. Its position on that axis can be found by integration provided that the equation of the curve used to generate the solid is known.

If the x-axis is the axis of symmetry, then $\bar{x}$, the x-coordinate of the C of M, is given by

$$\int \pi y^2 x \, dx = V\bar{x} \qquad \text{where } V \text{ is the volume of the solid}$$

Similarly, if the body is symmetrical about the y-axis,

$$\int \pi x^2 y \, dy = V\bar{y}$$

Quotable centres of mass are:

cylinder	midpoint of centre line
cone of height h	on axis of symmetry and $\frac{1}{4}h$ above the base,
pyramid of height h	on axis of symmetry and $\frac{1}{4}h$ above the base,
hemisphere of radius a	on radius of symmetry and $\frac{3}{8}a$ from the centre

Equilibrium of Rigid Bodies

Three conditions are necessary for the equilibrium of a three-dimensional object. One useful set of three is

the resultant force in a specified direction, Ox say, is zero,

the resultant force in the perpendicular direction, Oy, is zero,

the resultant moment about any one axis perpendicular to the plane is zero.

An alternative set which is sometimes more convenient to use is

the resultant force in a direction parallel to either Ox or Oy is zero,

the resultant moment about an axis through A is zero,

the resultant moment about another axis through B is zero.

Suspended Solids

When a body whose centre of mass is G is suspended from a point P, PG is vertical.

Bodies Resting on a Horizontal Plane

For equilibrium the vertical through G must pass through a point within the base of contact. If the surface of the body in contact with the plane is spherical, the weight must pass through the point of contact.

Bodies Resting on an Inclined Plane

As for a horizontal plane, the weight must pass through a point within the base of contact. In the case of a body with circular cross-section there is contact at only one point so the weight must pass through that point.

If no extra supporting force is applied to the body, there must be friction between the plane and the body in order to maintain equilibrium. This means that equilibrium depends not only upon the position of G relative to the contact region but also on whether slipping can occur.

MISCELLANEOUS EXERCISE B

1. A uniform ladder of length 5 m and weight 80 newton stands on rough level ground and rests in equilibrium against a smooth horizontal rail which is fixed 4 m vertically above the ground. If the inclination of the ladder to the *vertical* is θ, where $\tan\theta \leqslant \frac{3}{4}$, find expressions in terms of θ for the vertical reaction R of the ground, the friction F at the ground and the normal reaction N at the rail.

   Given that the ladder does not slip, show that F is a maximum when

   $\tan\theta = \dfrac{1}{\sqrt{2}}$, and give this maximum value.

   The coefficient of friction between the ladder and the ground is $\frac{1}{5}$. How much extra weight should be added at the bottom of the ladder so that the ladder will not slip when $\tan\theta = \frac{3}{4}$.   (OCSEB)

2.

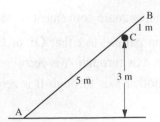

A smooth horizontal rail is fixed at a height of 3 m above a horizontal playground whose surface is rough. A straight uniform pole AB, of mass 20 kg and length 6 m, is placed to rest at a point C on the rail with the end A on the playground. The vertical plane containing the pole is at right angles to the rail. The distance AC is 5 m and the pole rests in limiting equilibrium.

Calculate

(a) the magnitude of the force exerted by the rail on the pole, giving your answer to the nearest N,

(b) the coefficient of friction between the pole and the playground, giving your answer to 2 decimal places,

(c) the magnitude of the force exerted by the playground on the pole, giving your answer to the nearest N.   (ULEAC)

3. The diagram shows a ladder AB, of mass 15 kg and length 8 m, whose lower end A rests on rough horizontal ground. The ladder is inclined at 60° to the horizontal supported by a smooth rail at C, where AC = 6 m. By modelling the ladder as a uniform rod, find

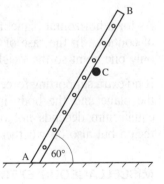

(a) the force exerted on the ladder by the rail at C,

(b) the vertical and the horizontal components of the force exerted on the ladder by the ground at A.

Given further that the ladder is in limiting equilibrium,

(c) find the coefficient of friction between the ladder and the ground.   (ULEAC)$_s$

In questions 4 to 8 a problem is set and is followed by a number of suggested responses. Choose the correct response.

For questions 4 and 5 use this diagram of a bead of mass m, threaded on to a smooth circular wire, of radius a, that is fixed in a vertical plane.

The bead is projected from the lowest point of the wire with speed $\sqrt{6ga}$.

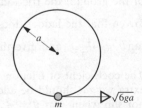

4. The normal reaction of the wire on the bead at the lowest point is

   A $6mg$ B $7mg$ C $8mg$ D $9mg$ E $10mg$

5. The bead reaches the highest point of the wire with a speed

   A 0   B $\sqrt{ga}$ C $\sqrt{2ga}$ D $\sqrt{3ga}$ E $\sqrt{4ga}$

6. A bead is threaded onto a circular wire fixed in a vertical plane. The bead
   travels freely round the wire. The acceleration of the bead is

   A   towards the centre and constant

   B   along the tangent and variable

   C   made up of two components one radial and one tangential

   D   away from the centre and variable.

7. A smooth hollow cylinder of radius a is fixed with its
   axis horizontal. A particle of mass m is projected from
   the lowest point of the inner surface with speed u.
   If $u = \sqrt{2ga}$ the particle will

   A   oscillate through 90°

   B   perform complete circles

   C   leave the cylinder

   D   oscillate through 180°.

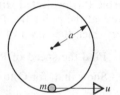

8. A particle of mass m, travelling in a vertical circle at the end of an inelastic string
   of length l, will describe complete circles provided that

   A   the kinetic energy at the lowest point exceeds $2mgl$

   B   the string never goes slack

   C   the potential energy at the highest point is greater than $2mgl$

   D   the tension in the string is constant.

9. A particle P of mass m lies inside a fixed smooth hollow sphere of internal radius
   a and centre O. When P is at rest at the lowest point A of the sphere, it is given
   a horizontal impulse of magnitude mu.

   The particle P loses contact with the inner surface of the sphere at the point B,
   where $\angle AOB = 120°$.

   (a) Show that $u^2 = \frac{7}{2}ga$.

   (b) Find the greatest height above B reached by P.   (ULEAC)$_s$

10. A stone, of mass 1.5 kg, tied to one end of a light inextensible string, is describing circles in a fixed vertical plane, the other end of the string being fixed. Given that the maximum speed of the stone is twice the minimum speed, prove that when the string is horizontal the magnitude of the tension in the string is 49 N. (AEB)

11.

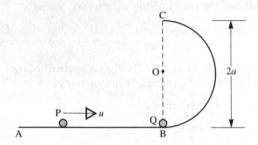

In a mechanics experiment, small marbles P and Q, of the same size but having masses 2m and m respectively, slide along a smooth groove in a rail.

The rail, which is fixed in a vertical plane, is in the form of a horizontal straight length AB joined to a semicircular arc BC of radius a and centre O, as shown in the diagram.

Marble P is projected with speed u along AB towards marble Q, which is initially at rest at the point B. There is a direct collision between P and Q, immediately after which the speed of Q is $\frac{5}{4}u$.

(a) Find the speed of P immediately after the collision.

(b) Show that, for Q to remain in contact with the groove on the complete arc BC,

$$u \geqslant \sqrt{\left(\frac{16}{5}ga\right)}.$$ (ULEAC)

12. A ball of mass m is travelling in a vertical plane round the inside of a circular track of radius 1 metre, as shown in the diagram.

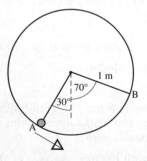

(a) What was its speed at the point A if it just reaches B before rolling back again?

(b) What must its minimum speed be at A if it is to complete the circle with out leaving the track? (SMP)$_s$

13.

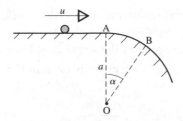

A toboggan of mass M, moving with speed u on smooth level ground, suddenly encounters a smooth downward slope, as shown in the diagram. The motion takes place in a vertical plane, and the downward slope is in the shape of an arc of a circle, radius a and centre O. The points A and B on the arc are such that OA is vertical and $\angle AOB = \alpha$, where $\cos \alpha > \frac{2}{3}$.

(a) Show that, for the toboggan to remain in contact with the slope on arc AB,

$$u^2 \leqslant ag(3 \cos \alpha - 2).$$

Given that $\cos \alpha = \frac{5}{6}$, and that the toboggan loses contact with the slope at the point B,

(b) find the sudden decrease, at the point A, in the magnitude of the force exerted by the ground on the toboggan. (ULEAC)

14. One section of a 'Loop the Loop' ride at an Adventure Park takes passengers round a vertical circle.

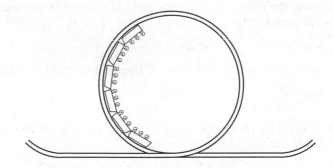

The situation is modelled by considering a small bead P of mass m threaded on a fixed smooth circular wire. The circular wire has centre O and radius a, and its plane is vertical. The bead is projected from the lowest point of the wire with speed $\sqrt{ag}$. When OP makes an angle θ with the downward vertical, find

(a) the speed of the bead.

(b) the reaction of the wire on the bead. (AEB)

15. A 'big wheel' at a fairground is rotating in a vertical circle at a constant rate of 1 revolution per minute. Take the value of g as 10.

  (i)   Calculate the angular speed of the wheel in radians per second.

 (ii)   Calculate the linear speed of a point 10 m from the axis of rotation.

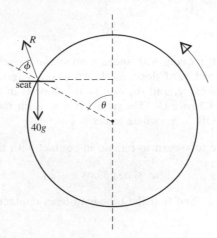

A child of mass 40 kg is sitting 10 m from the axis of rotation on a seat on the wheel. The situation may be modelled by assuming that there is negligible distance between the seat and the child's centre of mass and that the seat remains horizontal at all times. When the radius joining the seat to the axis is inclined at θ to the upward vertical, the reaction R of the seat on the child is inclined at ϕ to that radius, as shown in the diagram.

(iii)   By resolving the forces acting on the child in a direction perpendicular to the radius explain why

$$R \sin \phi = 40g \sin \theta.$$

(iv)   Write down the equation of motion of the child for the radial direction.

 (v)   Given that $\theta = \dfrac{\pi}{6}$, find R and ϕ.   (MEI)

In questions 16 and 17 a problem is set and is followed by a number of suggested responses. Choose the correct response.

16. A sphere A of mass $2m$ collides directly with a sphere B of mass m. Before impact each sphere is moving with speed u and A is brought to rest by the collision.

After impact the speed of B is

A $\frac{1}{2}u$ **B** u **C** $2u$ **D** $-u$

17. Two smooth objects, with a coefficient of restitution e, collide directly and bounce as shown

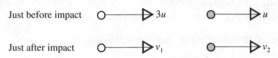

Newton's law of restitution gives

A $e \times 4u = v_2 + v_1$

B $e \times 2u = v_1 - v_2$

C $e \times 2u = v_2 - v_1$

D   it cannot be applied as the masses are not known.

In each question from 18 to 20 a statement is made. Decide whether the statement is true (T) or false (F).

18. A perfectly elastic impact does not cause a loss in mechanical energy.

19. The momentum of a system remains constant in any direction in which no external force acts.

20. Impulse means an impact between moving bodies.

21. In this question a situation is described and is followed by several statements. Decide whether each of the statements is true (T) or false (F).

A sphere A, of mass m, is moving with speed $2u$. It collides directly with another sphere B, also of mass m, which is initially at rest. The coefficient of restitution is $\frac{1}{2}$.

  (i)   After impact A's speed is $\frac{1}{2}u$.

 (ii)   There is no loss in kinetic energy at impact.

(iii)   The impulse that A exerts on B is twice the impulse that B exerts on A.

(iv)   After impact B's speed is greater than A's.

22. Two spheres A and B, of equal radius and of mass m and $5m$ respectively, are moving on the surface of a smooth horizontal table. The sphere A, moving with speed $6u$, strikes directly the sphere B, which is initially moving in the same direction as A with speed u. After the impact, the speed of A is $2u$ and its direction of motion is reversed.

(a)   Find the coefficient of restitution between A and B.

(b)   Show that the kinetic energy lost in the impact is $1.6mu^2$.   (ULEAC)$_s$

23. Two small smooth spheres A and B of mass $2m$ and $5m$ respectively, moving along Ox, collide. The velocity of A immediately before collision is $5u$ in the positive x direction, and immediately after collision the velocities of A and B in the positive x direction are $2u$ and $4u$ respectively. Determine

(i) the velocity of B immediately before collision,

(ii) the magnitude of the impulse on B,

(iii) the value of the coefficient of restitution. (WJEC)$_s$

24. A particle A of mass m, moving with speed u on a smooth horizontal surface, collides directly with a stationary particle B of mass $3m$.

The coefficient of restitution between A and B is e. The direction of motion of A is reversed by the collision.

(a) Show that the speed of B after the collision is $\frac{1}{4}u(1+e)$.

(b) Find the speed of A after the collision.

Subsequently, B hits a wall fixed at right angles to the direction of motion of A and B.

The coefficient of restitution between B and the wall is $\frac{1}{2}$. After B rebounds from the wall, there is another collision between A and B.

(c) Show that $\frac{1}{3} < e < \frac{3}{5}$.

(d) In the case $e = \frac{1}{2}$, find the magnitude of the impulse exerted on B by the wall. (ULEAC)

25. A small rubber ball is held at height h above a smooth level floor and released from rest at time $t = 0$. If the coefficient of restitution between the ball and the floor is e, show that after the first bounce the ball rises to a height h_1 where $h_1 = e^2 h$.

The ball continues to bounce until it comes to rest. Show that the total distance travelled by the ball from initial release to rest is $\dfrac{1 + e^2}{1 - e^2}\, h$.

Find

(i) the time when the ball first hits the floor,

(ii) the time between the first and second impacts of the ball on the floor.

Show that the ball comes to rest when

$$t = \frac{1+e}{1-e}\sqrt{\left(\frac{2h}{g}\right)}.$$ (OCSEB)

26. A ball is dropped from a height of $10\,\text{m}$ on to level ground. When it bounces, it leaves the ground with speed $10.5\,\text{m s}^{-1}$ travelling vertically upwards. The mass of the ball is $0.1\,\text{kg}$. Air resistance may be neglected.

  (i)   Find the speed with which the ball first hits the ground.

 (ii)   Show that the coefficient of restitution between the ball and the ground is $\frac{3}{4}$.

(iii)   Find the impulse on the ball when it first hits the ground.

(iv)   Find the energy lost by the ball on its first bounce.

 (v)   Explain why the ball leaves the ground after the second bounce with a speed of $14 \times \left(\frac{3}{4}\right)^2 \text{m s}^{-1}$.

This model for the behaviour of the ball breaks down when the speed of the ball at impact is less than $0.1\,\text{m s}^{-1}$.

(vi)   On which impact does the model break down?   (MEI)

27. The diagram shows the shape of a 'slide' for a children's playground. The section DE is straight and BCD is a circular arc of radius $5\,\text{m}$. C is the highest point of the arc and CD subtends an angle of $30°$ at the centre.

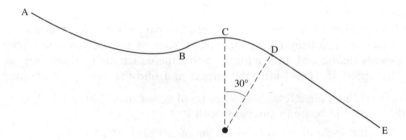

For safety reasons, children should not be sliding so fast that they lose contact with the slide at any point. Neglecting any resistances to motion, find

  (i)   the child's speed as it passes through C, given that it is on the point of losing contact at C,

 (ii)   the child's speed as it passes through D, given that it is on the point of losing contact as it reaches D.

Find the greatest possible height of the starting point A of the slide above the level of D, if a child starting rom rest at A is not to lose contact with the slide at any point.

Explain briefly whether taking resistances into account would lead to a larger or smaller value for the greatest 'safe height' above D.   (UCLES)$_s$

28. The diagram shows a smooth solid
hemisphere H of radius r fixed with its
plane surface, centre O, in contact with
horizontal ground. A particle is released
from rest on the surface of H at a point A
such that OA makes an angle α with the
upward vertical, where $\cos \alpha = \frac{7}{8}$. The
particle slides freely until it leaves the
surface of H at the point B with speed v.
Given that OB makes an angle α with the upward vertical,

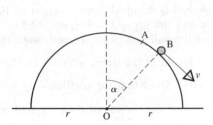

(a) show that $\cos \alpha = \frac{7}{12}$,

(b) find v^2 in terms of g and r.

Given that the particle strikes the ground with speed U,

(c) find U^2 in terms of g and r.   (ULEAC)

29. Two cars are being driven on a level 'skid pan', on which resistances to motion,
acceleration and braking may all be neglected.

Car A, of mass 1200 kg, is travelling at $15\,\text{m s}^{-1}$ when it collides directly with
car B, of mass 800 kg, travelling at $10\,\text{m s}^{-1}$ in the same direction.

(i) Let the speeds of cars A and B after the collision be v_A and v_B respectively.
Draw a diagram on which you mark the velocities of the cars before and
after collision.

(ii) The coefficient of restitution between the cars is 0.8. Write down
equations involving the momentum of each of the cars and their relative
speeds before and after impact. Solve the equations to show that car B
has speed $15.4\,\text{m s}^{-1}$ after the impact and find the new speed of car A.

Car B now collides directly with another small car of mass 740 kg which is
initially at rest and becomes entangled with it.

(iii) Find the speed of car B (and the entangled car) after this impact.

(iv) There is now a final direct collision between car A and car B (and the
entangled car) after which they separate at a speed of $2.55\,\text{m s}^{-1}$.
Calculate the coefficient of restitution in this impact.   (MEI)

In questions 30 to 32 a problem is set and is followed by a number of suggested
responses. Choose the correct response.

30. The diagram shows a solid cylinder with one of its plane
faces joined to the plane face of a solid cone. The
distance from V of the centre of mass of the combined
uniform solid is

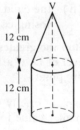

A   9 cm   B   12 cm   C   18 cm   D $15\frac{3}{4}$ cm

31. A uniform solid cone, of base radius a and height $4a$, rests with its flat face on an inclined plane that is rough enough to prevent slipping. The cone will be about to topple when

A $\alpha = 45°$ **B** $\tan \alpha = \frac{1}{4}$ **C** $\tan \alpha = \frac{3}{4}$

D $\alpha = 90°$ **E** $\tan \alpha = \frac{1}{2}$

32. A container consists of a hollow cylinder joined to a solid hemisphere as shown. When it is placed on a horizontal plane and tilted, it always returns to the position where AC is vertical. The centre of mass of the container is

A between B and C   **B** at A   **C** at B

D between B and A

33. In this question a situation is described and is followed by several statements. Decide whether each of the statements is true (T) or false (F).

A uniform solid body consists of a hemisphere and a cone joined together as shown. The centre of mass of the body is at O, the centre of the common plane face. When placed on an inclined plane, sufficiently rough to prevent slipping, the solid can rest in equilibrium on the plane in each of the following positions.

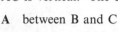

(i)   (ii)   (iii)   (iv)

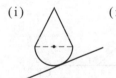  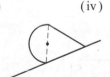

34. A uniform solid paperweight is in the shape of a frustum of a cone, as shown in the diagram. It is formed by removing a right circular cone of height h from a right circular cone of height $2h$ and base radius $2r$.

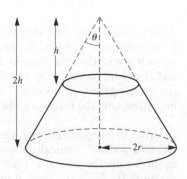

(a) Show that the centre of mass of the paperweight lies at a height of $\frac{11}{28}h$ from its base.

When placed with its curved surface on a horizontal plane, the paperweight is on the point of toppling.

(b) Find θ, the semi-vertical angle of the cone, to the nearest degree.

                                                       (ULEAC)

35. The diagram shows a man suspended by means of a
rope which is attached at one end to a peg at a
fixed point A on a vertical wall and at the other to
a belt round his waist. The man has weight $80g$ N,
the tension in the rope is T and the reaction of the
wall on the man is R. The rope is inclined at 35° to
the vertical and R is inclined at $\alpha°$ to the vertical as
shown. The man is in equilibrium.

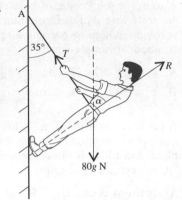

   (i)   Explain why $R > 0$.

  (ii)   By considering his horizontal and vertical
              equilibrium separately, obtain two
              equations connecting T, R and α.

 (iii)   Given that $\alpha = 45°$,   show that T is
              about 563 N and find R.

 (iv)   What is the magnitude and the direction of the force on the peg at A?

The peg at A is replaced by a smooth pulley. The rope is passed over the pulley
and tied to a hook at B directly below A. Calculate

  (v)   the new value of the tension in the rope section BA,

 (vi)   the magnitude of the force on the pulley at A.   (MEI)

36.

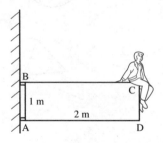

                  Fig. 1

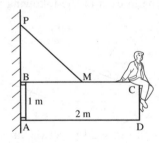

                  Fig. 2

A rectangular gate ABCD, where   AB $= 1$ m   and   AD $= 3$ m,   is supported
by smooth pins at A and B, where B is vertically above A. The pins are located
in such a way that the force at B is always horizontal. The gate has mass 120 kg
and it can be modelled by a uniform rectangular lamina. A boy, of mass 45 kg,
sits on the gate with his centre of mass vertically above C (see Fig. 1). Find the
magnitudes of the forces on the gate at B and at A.

In order to support the gate, the owner fits a cable attaching the mid-point M of
BC to a point P, vertically above B and such that   BP $= 1.5$ m. The boy once
again sits on the gate at C (see Fig. 2), and it is given that there is now no force
acting at B. Find the tension in the cable and the magnitude of the force now
acting at A.   (UCLES)$_s$

37. The diagram shows a sketch of the
region R bounded by the curve with
equation $y^2 = 4x$ and the line with
equation $x = 4$. The unit of length
on both the x-axis and the y-axis is
the centimetre. The region R is
rotated through π radians about the
x-axis to form a solid S.

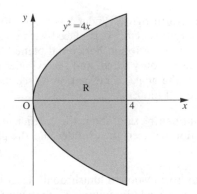

(a) Show that the volume of S is $32\pi\,\text{cm}^3$.

Given that the solid S is uniform,

(b) find the distance of the centre of
mass of S from O.   (ULEAC)

38.

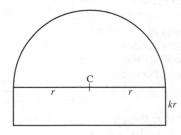

The diagram shows a cross-section containing the axis of symmetry of a uniform
body consisting of a solid right circular cylinder of base radius r and height kr
surmounted by a solid hemisphere of radius r. Given that the centre of mass of
the body is at the centre C of the common face of the cylinder and the
hemisphere, find the value of k, giving your answer to 2 significant figures.

Explain briefly why the body remains at rest when it is placed with any point of
its hemispherical surface in contact with a horizontal plane.   (ULEAC)

39. A mould for a right circular cone, base radius r
and height h, is produced by making a conical
hole in a uniform cylindrical block, base radius
$2r$ and height $3r$. The axis of symmetry of the
conical hole coincides with that of the cylinder,
and AB is a diameter of the top of the
cylinder, as shown in the diagram.

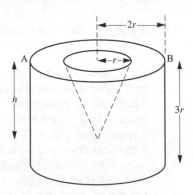

(a) Show that the distance from AB of
the centre of mass of the mould is

$$\frac{216r^2 - h^2}{4(36r - h)}.$$

The mould is suspended from the point A, and hangs freely in equilibrium.

(b) In the case $h = 2r$, calculate, to the nearest degree, the angle between AB
and the downward vertical..   (ULEAC)

40. A uniform right cylinder has height 40 cm and
base radius r cm. It is placed with its axis
vertical on a rough horizontal plane. The
plane is slowly tilted, and the cylinder topples
when the angle of inclination θ (see diagram)
is 20°. Find r.

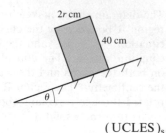
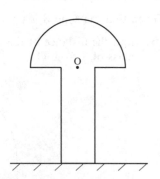

What can be said about the coefficient of
friction between the cylinder and the plane?

(UCLES)$_s$

41. A uniform wooden 'mushroom', used in a
game, is made by joining a solid cylinder to
a solid hemisphere. They are joined
symmetrically, such that the centre O of the
plane face of the hemisphere coincides with
the centre of one of the ends of the cylinder.
The diagram shows the cross-section
through a plane of symmetry of the
mushroom, as it stands on a horizontal
table.

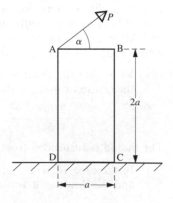

The radius of the cylinder is r, the radius of the hemisphere is $3r$, and the centre
of mass of the mushroom is at the point O.

(a) Show that the height of the cylinder is $r\sqrt{\left(\frac{81}{2}\right)}$.

The table top, which is rough enough to prevent the mushroom from sliding, is
slowly tilted until the mushroom is about to topple.

(b) Find, to the nearest degree, the angle with the horizontal through which the
table top has been tilted.   (ULEAC)

42. (i) An object is in equilibrium under the action of exactly three forces. What
must be true about the forces?

(ii) The figure shows a cross-section ABCD
of a uniform rectangular box of weight W.
The centre of mass of the box lies in the
plane ABCD and AB $= a$, BC $= 2a$.
The box rests with CD on a rough
horizontal floor and is pulled with
force P by a rope attached at A.
The rope is at an angle α to the
horizontal, as shown in the figure, and
the box is on the point of rotating
about C. Draw a diagram showing all
the forces acting on the box.

By taking moments about C, prove that $P = W/(2 \sin \alpha + 4 \cos \alpha)$.

Hence, or otherwise, show that P is a minimum when $\tan \alpha = \frac{1}{2}$. State this minimum value of P.

(iii)   When $\tan \alpha = \frac{1}{2}$ and the box is on the point of rotating about C without first sliding along the floor, show that μ, the coefficient of friction between the box and the floor, must be at least $\frac{2}{9}$.

What is the magnitude of the reaction at C?

(iv)   If $\mu = \frac{1}{5}$, what is the minimum value of P necessary to cause the box to rotate about C without sliding, and what is the corresponding value of α?

(OCSEB)

43.

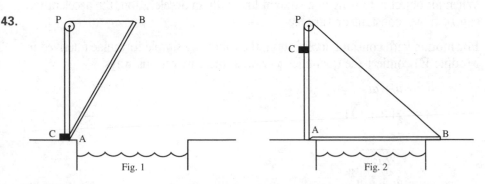

Fig. 1   Fig. 2

The diagrams show a simple mechanism by which a bridge over a canal is raised and lowered. The bridge AB is hinged at A, and a rope attached to B passes over a pulley P located vertically above A, at the top of a fixed vertical structure. A counterweight C is attached to the rope.

Figure 1 shows the bridge in the raised position, with C on the ground and B at the same horizontal level as P, and Fig. 2 shows the bridge lowered to its horizontal position. The mass of the bridge AB is 300 kg, and when raised (as in Fig. 1) the angle PAB is 30°.

Making suitable assumptions, which should be stated, find

(i)   the least mass needed for the counterweight C if it is to be capable of holding the bridge in the raised position,

(ii)   the extra force that needs to be applied to start raising the bridge from the horizontal position by pulling on the rope, if the mass of C is the minimum value found in (i).

(UCLES)ₛ

CHAPTER 8

VARIABLE ACCELERATION

SIMPLE MOTION IN A STRAIGHT LINE

When an object is moving in a straight line with an acceleration, the acceleration can be either constant or variable.

For motion with constant acceleration, the following simple formulae (derived in Module E) connect the quantities u, v, a, s and t in various ways.

$$v = u + at$$

$$s = \tfrac{1}{2}(u+v)t$$

$$s = \frac{v^2 - u^2}{2as}$$

$$s = ut + \tfrac{1}{2}at^2$$

$$s = vt - \tfrac{1}{2}at^2$$

It is vital to remember that these formulae are valid *only for motion with constant acceleration.*

When the acceleration is variable, it may sometimes depend in some way on the time interval during which motion has taken place. Acceleration that changes with time also is dealt with in Module E, where we used the following relationships

$$a = \frac{dv}{dt} \qquad \text{and} \qquad v = \frac{ds}{dt}$$

$$v = \int a\, dt \qquad \text{and} \qquad s = \int v\, dt$$

Remember that each of the derived functions given above is positive in the direction in which s increases, i.e.

In Module E we saw that these relationships are equally valid when a, v and s are given in vector form, i.e.

$$\mathbf{a} = \frac{d\mathbf{v}}{dt} \qquad \text{and} \qquad \mathbf{v} = \frac{d\mathbf{s}}{dt}$$

$$\mathbf{v} = \int \mathbf{a} \, dt \qquad \text{and} \qquad \mathbf{s} = \int \mathbf{v} \, dt$$

Note that r is often used as an alternative to s, particularly when using vector form, e.g. $\mathbf{r} = \int \mathbf{v} \, dt$

Before moving on the reader is recommended to revise these types of motion by working through the following exercise.

EXERCISE 8a

1. A particle is moving with constant acceleration $5 \, \text{m s}^{-2}$ along a straight line. It started at a point O on the line with initial velocity $3 \, \text{m s}^{-1}$.

   (a) Find its velocity after 8 s.
   (b) Find its velocity after travelling 4 m.
   (c) Find its distance from O after 4 s.
   (d) Find its distance from O when its velocity is $13 \, \text{m s}^{-1}$.

2. A stone is projected vertically upwards at $30 \, \text{m s}^{-1}$ from a point O, just overhanging the edge of a cliff. After t seconds the upward displacement of the stone from O is s metres and its velocity is $v \, \text{m s}^{-1}$. Take $g = 10$.

   (a) Find v in terms of t.
   (b) Find s in terms of t.
   (c) Find the time when $v = 0$ and the height of the stone above O at this time.
   (d) Find the time when the stone passes the cliff top during its descent.
   (e) Find the length of time for which the stone is more than 40 m above the cliff top.
   (f) Find s and v when $t = 7$.
   (g) How long is it before the stone reaches sea level, 80 m below the cliff top?
   (h) Find the speed of the stone when it is 35 m below the cliff top.
   (i) Sketch graphs of s against t and v against t for $0 \leqslant t \leqslant 8$.

In questions 3 to 5, a particle P is moving on a straight line and O is a fixed point on that line. After a time of t seconds the displacement of P from O is s metres, the velocity is $v\,\mathrm{m\,s^{-1}}$ and the acceleration is $a\,\mathrm{m\,s^{-2}}$.

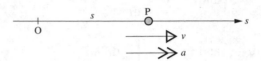

3. $s = t^3 - 9t^2 + 24t$

    (a) Find v in terms of t.

    (b) Find the initial velocity.

    (c) Find the times when P is instantaneously at rest and its displacement from O at these times.

    (d) Find a in terms of t.

    (e) Find (i) the time when the acceleration is zero,
                (ii) the velocity of this time,

    (f) Find the minimum velocity and the minimum speed.

4. $a = \dfrac{8}{(1+t)^3}$ and, when $t = 0$, $v = 8$ and $s = 4$.

    (a) Find v in terms of t.

    (b) Find s in terms of t.

    (c) Find the value which the velocity approaches as t increases.

5. $v = 60t - 6t^2$ and $s = 0$ when $t = 0$.

    (a) Find a in terms of t.

    (b) Find the maximum velocity.

    (c) Find the maximum speed in the interval $0 \leqslant t \leqslant 15$.

    (d) Find s in terms of t.

    (e) Find the time after the start when P is instantaneously at rest and the distance travelled up to this time.

    (f) Find the time when P again passes through O, its velocity at this time and the distance it has travelled up to this time.

    (g) Sketch the graph of v against t, for $0 \leqslant t \leqslant 15$, and indicate how the distances found in parts (e) and (f) are connected with this graph.

In questions 6 and 7 a particle P is moving in the xy plane and O is the origin. Unit vectors in the directions Ox and Oy are **i** and **j** respectively. For P the position vector relative to O at time t is **r**, the velocity vector is **v** and the acceleration vector is **a**. All units are consistent and based on metres and seconds.

6. $\mathbf{r} = (2t-1)\mathbf{i} + t^3\mathbf{j}$

   (a) Find **v** in terms of t.

   (b) Find **a** in terms of t.

   (c) Find the initial velocity.

   (d) When $t = 2$ find
      (i) the distance OP,
      (ii) the speed of P,
      (iii) the angle between the direction of motion of P and the x-axis.

   (e) Find the cartesian equation of the path of P.

7. $\mathbf{a} = \dfrac{32}{t^3}\mathbf{j}$ and, when $t = 2$, $\mathbf{v} = 9\mathbf{i} - 4\mathbf{j}$ and $\mathbf{r} = 18\mathbf{i} + 8\mathbf{j}$.

   (a) Find **v** in terms of t.

   (b) Find **r** in terms of t.

   (c) Find the value of t when **v** is perpendicular to **r**.

   (d) Show that, as t increases, **v** approaches a constant value and state the magnitude and direction of this terminal velocity as $t \to \infty$.

USING DIFFERENTIAL EQUATIONS

The problems in Exercise 8a were solved by interpreting given information about motion by means of a differential equation in which velocity and acceleration were denoted by derived functions such as $\dfrac{ds}{dt}$ and $\dfrac{dv}{dt}$.

These particular derived functions were suitable because both s and v were functions of t and the resulting equation therefore contained only two variables (either s and t, or v and t).

In practice the acceleration of a moving object is more likely to depend on the distance or speed rather than the time of motion. For instance, the tension in an elastic string depends on the extension, so the force acting on a particle attached to the end, and hence the acceleration of that particle, depends on its displacement from the natural length position. In such cases, where the acceleration is not a function of time, we may have to find an alternative derived function to represent acceleration.

ACCELERATION AS A FUNCTION OF DISPLACEMENT

The general case of motion with an acceleration that is a function of displacement is represented by $a = f(s)$.

The expression so far used for acceleration is $\dfrac{dv}{dt}$, but when acceleration is a function of s this gives $\dfrac{dv}{dt} = f(s)$.

There are three variables in this differential equation, v, s and t, so at this stage the equation cannot be solved and we must look for another way of expressing the acceleration.

The chain rule gives $\dfrac{dy}{dx} = \dfrac{dy}{du} \times \dfrac{du}{dx}$ where u is a third variable.

This relationship can be expressed in terms of the variables v, s and t,

i.e. $\qquad \dfrac{dv}{dt} = \dfrac{dv}{ds} \times \dfrac{ds}{dt}$

Now $\qquad \dfrac{ds}{dt} = v \qquad$ therefore $\qquad \dfrac{dv}{dt} = v \dfrac{dv}{ds}$

i.e. $\qquad\qquad\qquad$ acceleration $= v \dfrac{dv}{ds}$

Hence, when $a = f(s)$ we have $f(s) = v \dfrac{dv}{ds}$

Separating the variables in this differential gives

$$\int f(s) \; ds = \int v \; dv$$

If $f(s)$ is known (and can be integrated) a solution can be found.

Examples 8b

1. A particle P is moving along a straight line with an acceleration that is proportional to s^2 where s metres is the displacement of P from a fixed point A on the line.

   (a) Find a general relationship between the displacement and the velocity, v metres per second.

   Given that v and s are equal in magnitude when $s = 0$ and when $s = 4$,

   (b) find the speed when the displacement is 1.5 m

   (c) find the displacement when the velocity is $2 \, \text{m s}^{-1}$

   (d) sketch the graph of velocity against displacement for $0 \leqslant s \leqslant 4$.

(a) Acceleration $\propto s^2$

$$\therefore \qquad a = ks^2 \quad \text{where } k \text{ is a constant}$$

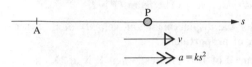

Using $\qquad a = v\dfrac{\mathrm{d}v}{\mathrm{d}s}$ gives

$$v\frac{\mathrm{d}v}{\mathrm{d}s} = ks^2 \qquad \Rightarrow \qquad \int v \, \mathrm{d}v = \int ks^2 \, \mathrm{d}s$$

$$\therefore \qquad \tfrac{1}{2}v^2 = \tfrac{1}{3}ks^3 + K$$

This relationship between s and v is general because it contains two unknown constants.

$$v = 0 \quad \text{when} \quad s = 0 \qquad \Rightarrow \qquad K = 0$$

$$\therefore \qquad\qquad\qquad\qquad\qquad 3v^2 = 2ks^3$$

$$v = 4 \quad \text{when} \quad s = 4 \qquad \Rightarrow \qquad 48 = 128k \qquad \Rightarrow \qquad k = \tfrac{3}{8}$$

$$\therefore \qquad\qquad\qquad\qquad\qquad 4v^2 = s^3$$

(b) When $\quad s = 1.5, \qquad 4v^2 = (1.5)^3$

$$\Rightarrow \qquad\qquad\qquad\qquad v^2 = 0.843\ldots \qquad \Rightarrow \qquad v = \pm 0.918\ldots$$

The speed is $0.92\,\mathrm{m\,s^{-1}}$ (2 dp)

(c) When $\quad v = 2, \qquad 16 = s^3 \qquad \Rightarrow \qquad s = 2.519\ldots$

The displacement is $2.52\,\mathrm{m}$ (2 dp)

(d) Because $v = \pm\tfrac{1}{2}\sqrt{s^3}$, there are two values of v for each value of s.

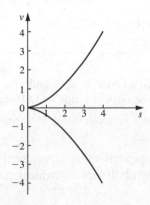

2. A particle P is moving in a straight line and O is a fixed point on the line. The magnitude of the acceleration of P is proportional to the distance of P from O; the direction of the acceleration is always towards O. Initially the particle is at rest at a point A where the *displacement* of A from O is *l*.

(a) Using x and v for the displacement and velocity of P, express v^2 in terms of x, l and a constant of proportion, k.

(b) Show that the speed of P is greatest when P is at O.

(c) Determine a position other than A where the velocity of P is zero.

(d) Describe briefly the motion of P.

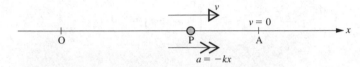

In the diagram the acceleration of P is towards O, i.e. in the negative direction, so $a = -kx$. If P is left of O, the acceleration is in the positive direction but x is negative, so again we have $a = -kx$.

(a) $a = -kx$

Using $a = v\dfrac{\mathrm{d}v}{\mathrm{d}x}$, gives $v\dfrac{\mathrm{d}v}{\mathrm{d}x} = -kx$

$\Rightarrow$ $\displaystyle\int v\ \mathrm{d}v = -\int kx\ \mathrm{d}x$

$\Rightarrow$ $\frac{1}{2}v^2 = -\frac{1}{2}kx^2 + K$

$v = 0$ when $x = l$ $\Rightarrow$ $K = \frac{1}{2}kl^2$

$$v^2 = k(l^2 - x^2)$$

(b) The value of x^2 can never be negative, so the greatest value of $(l^2 - x^2)$ is $(l^2 - 0)$

The greatest speed occurs when $x = 0$, i.e. when P is at O.

(c) When $v = 0$, $k(l^2 - x^2) = 0$ $\Rightarrow$ $l^2 = x^2$

$\Rightarrow$ $x = \pm l$

i.e. the displacement of P from O is $\pm l$.

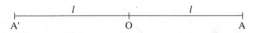

$\therefore$ the speed of P is zero when P is at A and at a point A′ where OA′ = l and A′ is on the opposite side of O from A.

(d) Wherever P is on the line it is accelerating towards O, so P oscillates between A and A′.

Note that in Example 2, displacement is represented by the symbol x as an alternative to s. The two symbols are equivalent and either may appear in questions.

EXERCISE 8b

In questions 1 to 10 a particle P is moving along a straight line and O is a fixed point on that line. After a time of t seconds the displacement of P from O is s metres, the velocity is $v\,\mathrm{m\,s^{-1}}$ and the acceleration is $a\,\mathrm{m\,s^{-2}}$.

1. $a = 2s + 5$ and $v = 2$ when $s = 0$.

(a) Find v^2 in terms of s.

(b) Find the speed when $s = 1$.

(c) Find s when $v = 4$.

2. $a = 6s + 4$ and $v = 3$ when $s = 0$.

(a) Find v^2 in terms of s.

(b) Find the speed when $s = 2$.

(c) Find s when $v = 5$.

3. $a = s - 4$ and $v = 12$ when $s = 4$.

(a) Find v^2 in terms of s.

(b) Find the speed when $s = 0$ and when $s = 9$.

(c) For what value of s is $\dfrac{dv}{ds}$ zero?

(d) Sketch the graph of v against s for $0 \leqslant s \leqslant 9$.

4. P is moving in the positive direction with an acceleration given by $a = 8s^3$, and when $s = 0$, $v = 0$.

(a) Find v in terms of s.

(b) Find the value of s for which $v = 50$.

5. P is moving in the positive direction with an acceleration proportional to s^3. When $s = 0$, $v = 0$ and when $s = 1$, $v = 9$.

(a) Find v in terms of s.

(b) Express a in terms of s.

(c) Find the value of s for which $v = 900$.

6. When $v = 7$, $s = 5$ and when $v = 18$, $s = 6$. Given that a is proportional to s and $s > 0$ for all values of t,

(a) find v^2 in terms of s

(b) find the value of s when $v = 10$

(c) find the least distance of P from the origin.

7. $a = 36 - 12s^2$ and $v = 0$ when $s = 0$.

   (a) Find v^2 in terms of s.

   (b) Find the speed when $s = 1$.

   (c) Find all the values of s for which $v = 0$.

   (d) By considering the sign of v^2, show that motion can take place on the section of the line for which $0 \leqslant s \leqslant 3$.

   (e) Find the acceleration   (i) when $s = 0$ (ii) when $s = 3$.

   (f) Find the value of s when v is a maximum and find the maximum speed.

   (g) Describe the motion of P.

8. $a = e^s$ and $v = 2$ when $s = 0$.

   (a) Find v^2 in terms of s.

   (b) Find v when $s = 4$.

   (c) Find s when $v = 20$.

9. $a = 40e^{-s}$ and $v = 8$ when $s = 0$.

   (a) Find v^2 in terms of s.

   (b) Find s when $v = 11$.

   (c) Find the terminal speed, i.e. the speed which is approached as s increases indefinitely.

10. $a = \dfrac{10}{s+1}$ for $s \geqslant 0$, and, when $s = 0$, $v = 4$.

    (a) Find v^2 in terms of s.

    (b) Find v when $s = 4$.

    (c) Find s when $v = 12$.

11. A particle moves in a straight line with an acceleration $12s^2 \, \text{m s}^{-2}$ where s metres is the displacement of the particle from O, a fixed point on the line, after t seconds. The particle has zero velocity when its displacement from O is $-2 \, \text{m}$. Find the velocity of the particle as it passes through O.

12. A spacecraft is moving in a straight line directly away from the centre of the earth. When it is at a distance x kilometres from the centre of the earth its acceleration, which is due to gravity, is $\dfrac{4 \times 10^5}{x^2} \, \text{km s}^{-2}$ towards the centre of the earth. When it is at 8000 km from the centre of the earth its speed is $11 \, \text{km s}^{-1}$.

    (a) Find v^2 in terms of x.

    (b) Find v when $x = 10\,000$.

    (c) Find the terminal velocity.

VELOCITY GIVEN AS A FUNCTION OF DISPLACEMENT

In the previous section we saw that when acceleration is expressed in terms of displacement, using $v\dfrac{\mathrm{d}v}{\mathrm{d}s}$ for the acceleration results in a relationship between velocity and displacement.

So if the motion of a particle is defined by $v = f(s)$, it seems logical that we should be able to reverse the process to find the acceleration from this relationship.

Finding Acceleration as a Function of Displacement

If $v = f(s)$ and we want to find the acceleration, using $a = \dfrac{\mathrm{d}v}{\mathrm{d}t}$ is of no help as we cannot differentiate $f(s)$ with respect to t.

However, as we saw earlier in the chapter, $\dfrac{\mathrm{d}v}{\mathrm{d}t} \equiv v\dfrac{\mathrm{d}v}{\mathrm{d}s}$, and this form for the acceleration is useful again here.

Examples 8c

1. The velocity, in metres per second, of a particle P moving in a straight line is given by $v = x + \dfrac{1}{x}$ where x metres is the displacement of P from a fixed point O on the line. Find the acceleration of P when $x = 2$

$$v = x + \frac{1}{x} \quad \Rightarrow \quad \frac{\mathrm{d}v}{\mathrm{d}x} = 1 - \frac{1}{x^2}$$

Using $a = v\dfrac{\mathrm{d}v}{\mathrm{d}x}$ gives $a = \left(x + \dfrac{1}{x}\right)\left(1 - \dfrac{1}{x^2}\right)$

$$\Rightarrow \qquad a = x - \frac{1}{x^3}$$

When $x = 2$, $a = 2 - \frac{1}{8} = 1\frac{7}{8}$

When $x = 2$ the acceleration is $1.875\,\mathrm{m\,s^{-2}}$.

Relating Displacement and Time

For motion defined by $v = f(s)$, in order to relate displacement and time we can use $v = \dfrac{ds}{dt}$, giving $\dfrac{ds}{dt} = f(s)$.

This differential equation contains only two variables so it provides a suitable method for trying to find a relationship between s and t.

Consider a particle, travelling in a straight line, whose velocity v is given by $v = \sqrt{s}$, where s is the displacement of the particle from a fixed point on the line.

Using $v = \dfrac{ds}{dt}$ gives $\dfrac{ds}{dt} = s^{\frac{1}{2}}$

$\Rightarrow \qquad\qquad\qquad \int s^{-\frac{1}{2}} \, ds = \int dt \quad \left(\text{i.e.} \int 1 \, dt \right)$

$\Rightarrow \qquad\qquad\qquad 2s^{\frac{1}{2}} = t + K$

To find the exact relationship between s and t, further information is needed to give the value of K.

In general, if $v = f(s)$ we have

$$\frac{ds}{dt} = f(s) \qquad \Rightarrow \qquad \int \frac{1}{f(s)} \, ds = \int dt$$

Whether we can go any further depends on whether it is possible to perform the integration on the left hand side of this equation. In Example 8b/2, for example, from which we find that $v = \sqrt{k(l^2 - x^2)}$, many readers would be unable to carry on because they have not yet seen a method for integrating $\dfrac{1}{\sqrt{(l^2 - x^2)}}$ with respect to x.

In this book, however, any question in which a type of motion is defined by a differential equation will lead to an expression that can be integrated at this stage.

Examples 8c (continued)

2. The velocity, $v\,\mathrm{m\,s^{-2}}$, of a particle P at any time t is proportional to the square of the displacement, s metres, of P from a fixed point A. P is moving in a straight line through A. Initially, i.e. when $t = 0$, P is 2 m from A and, 3 seconds later, $\mathrm{AP} = 1.25\,\mathrm{m}$.

(a) Find s when $t = 2$

(b) Show that, as the time increases indefinitely, the displacement of P from A approaches a particular value and state the position that P is then approaching.

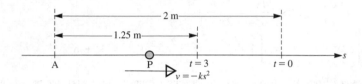

In this example it is clear from the positions of P when $t = 0$ and $t = 3$, that P is travelling towards O, i.e. v is negative.

$$\therefore \qquad v \propto s^2 \qquad \Rightarrow \qquad v = -ks^2$$

Using $v = \dfrac{ds}{dt}$ gives $\dfrac{ds}{dt} = -ks^2$

Hence $\qquad -\displaystyle\int \frac{1}{s^2}\,ds = \int k\ dt \qquad$ i.e. $\quad -\displaystyle\int s^{-2}\,ds = \int k\ dt$

$$s^{-1} = kt + K$$

When $t = 0$, $s = 2$, therefore $K = 0.5$

$\Rightarrow \qquad\qquad\qquad s^{-1} = kt + 0.5$

When $t = 3$, $s = 1.25$, therefore $0.8 = 3k + 0.5 \qquad \Rightarrow \qquad k = 0.1$

$\therefore \qquad\qquad\qquad s^{-1} = 0.1t + 0.5$

(a) When $t = 2$, $\quad s^{-1} = 0.2 + 0.5 = 0.7 \quad \Rightarrow \quad s = 1.43 \quad (3\ \mathrm{sf})$

(b) First we will express s as a simplified function of t.

$$s^{-1} = 0.1t + 0.5 \qquad \Rightarrow \qquad 10s^{-1} = t + 5$$

$$\therefore \qquad\qquad s = \frac{10}{t+5}$$

As t becomes very large, s becomes very small, i.e. s approaches 0. When t is very large, P approaches A.

Note that it was not necessary to notice that v was negative; the calculation of the constant k sorts out the correct sign.

3. A particle is travelling along the line ABC as shown in the diagram.

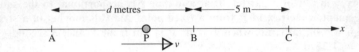

The velocity, $v \, \text{m s}^{-1}$, is given by $\quad v = \dfrac{4}{2x+1} \quad$ when the particle is x m from A.

Given that the particle travels from B to C, a distance of 5 m, in 12 seconds, find the distance from A to B.

Using $\quad v = \dfrac{\mathrm{d}x}{\mathrm{d}t} \quad$ gives $\quad \dfrac{\mathrm{d}x}{\mathrm{d}t} = \dfrac{4}{2x+1}$

$\therefore \qquad \displaystyle\int (2x+1) \; \mathrm{d}x = \int 4 \; \mathrm{d}t \qquad \Rightarrow \qquad x^2 + x = 4t + K$

As x is measured from A, it is convenient to measure t from the time when the particle is at A, so that $x = 0$ when $t = 0$.

When the particle is at A, $\quad x = 0 \quad$ and $\quad t = 0 \qquad \Rightarrow \qquad K = 0$

$\therefore \qquad 4t = x^2 + x$

Let the distance AB be d m and the time taken to travel from A to B be T seconds.

When the particle is at B, $\quad x = d \quad$ and $\quad t = T$

$\therefore \qquad 4T = d^2 + d$ [1]

When the particle is at C, $\quad x = (d+5) \quad$ and $\quad t = (T+12)$

$\therefore \qquad 4(T+12) = (d+5)^2 + (d+5)$

$\Rightarrow \qquad 4T + 48 = d^2 + 10d + 25 + d + 5$

$\Rightarrow \qquad 4T = d^2 + 11d - 18$ [2]

$[2] - [1]$ gives $\quad 0 = 10d - 18$

$\therefore \qquad\qquad d = 1.8$

The distance from A to B is 1.8 m.

EXERCISE 8c

In questions 1 to 6 a particle P is moving along a straight line and O is a fixed point on that line. After a time of t seconds the displacement of P from O is s metres, the velocity is $v \, \text{m s}^{-1}$ and the acceleration in $a \, \text{m s}^{-2}$.

1. If $\quad v = \dfrac{1}{s^2}, \quad$ find

(a) a in terms of s (b) a when $s = 0.5$.

2. Given that v is proportional to s, show that a is also proportional to s.

   Given that $v = 6$ when $s = 2$, find a when $s = 2$.

3. Given that $v = \dfrac{5}{1 + 2s}$ find a in terms of s, and the value of a when $s = 2$.

4. Given that $v = \sqrt{s}$ and $s = 0$ when $t = 0$, find

   (a) s in terms of t

   (b) the times taken to travel   (i) the first 25 m   (ii) the next 25 m

   (c) v when $t = 5$.

5. If $s = 5$ when $t = 0$, and $v = -\frac{1}{4}s$, find

   (a) s in terms of t

   (b) s when $t = 2$

   (c) the value approached by s as t increases.

6. The table shows corresponding values of v and s.

s	0	1	3
v	12	10	6

   (a) Show that all these values fit a relationship of the form $v = cs + d$ and state the values of c and d.

   (b) Show that $\ln(6 - s) = A - 2t$, where A is a constant.

   (c) Given that $s = 0$ when $t = 0$, express s in terms of t.

   (d) Find the maximum distance of P from O.

7. A body is moving on a horizontal straight line through a liquid. It passes through a fixed point O on the line and t seconds later its displacement from that point is s metres, its velocity is $v\,\mathrm{m\,s^{-1}}$ and its acceleration is $a\,\mathrm{m\,s^{-2}}$. The motion of the body is modelled by the equation

   $$v = pe^{ks}, \quad \text{where } p \text{ and } k \text{ are constants.}$$

   (a) Show that this is consistent with the hypothesis that the acceleration is given by $a = kv^2$.

   (b) Find t in terms of s, p and k given that, at $t = 0$, $s = 3$.

   (c) Find p and k given that $v = 20$ when $s = 0$ and $v = 10$ when $s = 3$.

   (d) Find the *distance* travelled when the velocity has fallen to $5\,\mathrm{m\,s^{-1}}$.

8. A particle P is moving along a straight line and O is a fixed point on that line. After t seconds the displacement of P from O is s m and P's velocity is v m s^{-1}. Throughout the motion v is inversely proportional to s and $s > 0$.
   When $t = 0$, $s = 4$ and P takes 7 seconds to move from the point where $s = 6$ to the point where $s = 8$. Find

   (a) s in terms of t (b) v in terms of s (c) v when $t = 2$.

9. The motion of a body falling vertically through a liquid is modelled by the equation

$$v^2 = \frac{g}{k}\left(1 - e^{-2ks}\right)$$

   where, at time t seconds, v m s^{-1} is its velocity and s metres is its displacement from its initial position and k is a constant.

   (a) It is observed that, after falling several metres, its velocity starts to approach a constant value which is estimated to be 7 m s^{-1}.
       Taking $g = 9.8$, find the value of k.

   (b) Find the acceleration in terms of s.

   (c) Show that the acceleration can be written in the form $a = g - 0.2v^2$.

*10. A cyclist approaches a hill at a speed of 9 m s^{-1} but slows down gradually as he climbs it. After climbing for t seconds his displacement from a point O at the bottom of the hill is s metres and his speed is v m s^{-1}. He thinks that his velocity may possibly be modelled by one of the following equations, in which λ and μ are constants.

   Model 1 $v = 9 - \lambda s$

   Model 2 $v = 9 - \mu^2 s^2$

   (a) Show that model 1 gives $t = \dfrac{1}{\lambda} \ln\left\{\dfrac{9}{9 - \lambda s}\right\}$

   (b) Show that model 2 gives $t = \dfrac{1}{6\mu} \ln\left\{\dfrac{3 + \mu s}{3 - \mu s}\right\}$

   (c) When he reaches the top of the hill, which is 100 m long, his speed has dropped to 1 m s^{-1}. Find the values of λ and μ.

   (d) As one check on whether either of these models is suitable he measures the time taken over the first 50 m and finds that it is 7 seconds. For each model find the time predicted for the cyclist to cover the 50 m. Does either model give a result consistent with the measured value?

   (e) He reaches the top of the hill in 27 seconds. Does this time strengthen the case for either model?

CHOOSING SUITABLE DERIVED FUNCTIONS

When attempting a solution to a problem on variable motion, it is important to represent velocity and/or acceleration in a way that leads to a differential equation that can be solved, i.e. not more than two variables appear in the equation. A summary of suitable differential equations for most situations is given below.

When a, v or s is a function of time, $f(t)$, use:

$$v = \frac{ds}{dt} \qquad a = \frac{dv}{dt}$$

$$s = \int v \ dt \qquad v = \int a \ dt$$

When a is a function of displacement, $f(s)$, use

$$a = v \frac{dv}{ds} \quad \Rightarrow \quad \int f(s) \ ds = \int v \ dv \quad (\text{giving } v \text{ as a function of } s)$$

When v is a function of displacement, $f(s)$, use

$$a = v \frac{dv}{ds} = f(s) \times \frac{d}{ds} f(s)$$

$$v = \frac{ds}{dt} = f(s) \quad \Rightarrow \quad \int \frac{1}{f(s)} \ ds = \int dt$$

Remember that these relationships apply when a, v and s (or r) are given in cartesian vector form.

An opportunity to practise making the best choice is given in the following exercise of mixed questions.

The Dot Notation

Note that an alternative notation can also be useful. In this notation a dot over a symbol means 'the derivative with respect to t of that quantity' i.e.

$$\dot{v} \equiv \frac{dv}{dt} = a \quad \text{and} \quad \dot{s} \equiv \frac{ds}{dt} = v$$

Further, as $a = \frac{d}{dt}\left(\frac{ds}{dt}\right) = \frac{d^2s}{dt^2}$, it follows that a can be denoted by $\ddot{s}$.

EXERCISE 8d

In questions 1 to 5 a particle P is moving along a straight line and O is a fixed point on that line. After a time of t seconds the displacement of P from O is s metres, the velocity is $v\,\mathrm{m\,s^{-1}}$ and the acceleration is $a\,\mathrm{m\,s^{-2}}$.

1. The motion satisfies the equation $v = \dfrac{k}{s^2}$. Further, $s = 2$ when $t = 1$ and $s = 3$ when $t = 10.5$.

    (a) Find s in terms of t.   (b) Find t when $s = 4$.

2. If $v = 8 - 6e^{-2t}$ and $s = 0$ when $t = 0$, find

    (a) s in terms of t (b) s when $v = 7.5$

3. The motion is described by the equation $v = -12 \sin 3t$

    (a) Find a in terms of t.
    (b) Find a when $t = 0.5$.
    (c) Find s in terms of t given that $s = 4$ when $t = 0$.
    (d) Show that $a = -n^2 s$ where n is a constant and state the value of n.

4. (a) If $a = 10t - 40$, for $0 \leqslant t \leqslant 10$, and $v = 0$ when $t = 0$,
        (i) find v in terms of t
        (ii) find t when the velocity is zero
        (iii) find the greatest speed in the positive direction and the greatest speed in the negative direction and the times at which they occur
        (iv) sketch a graph of v against t for $t \geqslant 0$.

    (b) If, instead, $a = 40 - 10t$ for $0 \leqslant t \leqslant 10$, sketch a graph of v against t for $t \geqslant 0$.

5. Given that $a = 8 - 2t^2$ and, when $t = 0, v = 0$ and $s = 0$, find

    (a) the greatest speed in the positive direction
    (b) the distance covered by the particle in the first two seconds of its motion.

In questions 6 to 8, a particle P is moving in the xy plane. O is the origin and, at any time t, the displacement of P from O is $\mathbf{r}$, the velocity vector is $\mathbf{v}$ and the acceleration vector is $\mathbf{a}$. All units are consistent and based on metres and seconds.

6. Given that $\mathbf{v} = 3t^2\mathbf{i} - 4t\mathbf{j}$

    (a) find $\mathbf{a}$ by differentiating $\mathbf{v}$ with respect to time
    (b) show that $\mathbf{r} = t^3\mathbf{i} - 2t^2\mathbf{j} + \mathbf{R}$, where $\mathbf{R}$ is a constant vector.
    If also $\mathbf{r} = 8\mathbf{i}$ when $t = 2$, find $\mathbf{R}$.

7. When $\mathbf{r} = e^{2t}\mathbf{i} + t^2\mathbf{j}$, find

   (a) $\mathbf{v}$ in terms of t

   (b) $\mathbf{a}$ in terms of t

   (c) the initial velocity and acceleration

   (d) the time at which $\mathbf{a} = 20\mathbf{i} + 2\mathbf{j}$ and the velocity at this time.

8. If $\mathbf{v} = 10\cos 2t\,\mathbf{i} - 10\sin 2t\,\mathbf{j}$ and $\mathbf{r} = 5\mathbf{j}$ when $t = 0$

   (a) find $\mathbf{r}$ in terms of t

   (b) find $\mathbf{a}$ in terms of t

   (c) find $\mathbf{r}$ when $t = \frac{\pi}{4}$

   (d) show that $|\mathbf{r}|$ is constant and state its value

   (e) show that $|\mathbf{v}|$ is constant and state its value

   (f) show that $\mathbf{a} = \lambda\mathbf{r}$ and state the value of λ

   (g) find the cartesian equation of the path of P.

9. After a car has travelled from rest along a straight level road, its acceleration, $a\,\mathrm{m\,s^{-1}}$, is given by $a = \dfrac{200 - s}{60}$ where s is the distance travelled in metres.

   (a) Find the speed of the car after it has travelled 25 metres.

   (b) Find the maximum speed it achieves and the distance it has then travelled.

   (c) How far from the starting point is it when it comes to rest again?

*10. An automatic conveyor is designed for a factory. It moves forwards and backwards along a straight track between two points A and B which are 81 m apart. It comes to instantaneous rest at A and B, and the journey between these points takes 6 seconds. A mathematical model for the twelve-second return journey, from A to B and back to A, is to be considered.

   Take an origin at A. Let the displacement t seconds after leaving A be s metres and the velocity be v metres per second.
   You are to consider whether the motion can be modelled by an equation of the form $v = kt(t - 6)(t - 12)$, for $0 \leqslant t \leqslant 12$, where k is a constant.

   (a) Verify that this is consistent with the given information about times and velocities.

   (b) Find s in terms of k and t and use the given information about the distance AB to find the value of k.

   (c) The model will not be suitable unless it gives $s = 0$ when $t = 12$. Is it suitable in this respect?

   (d) Find the maximum speeds predicted by the model on the outward and return journeys and the times at which these occur.

SIMPLE HARMONIC MOTION

One particular type of variable acceleration is important in its own right because it occurs frequently in everyday life. This is motion in a straight line in which the acceleration is proportional to the distance from a fixed point on the line, and is always directed towards that point; it is called *simple harmonic motion, SHM*. A weight attached to the end of a spring, for example, moves in this way; this, and other real-life cases of SHM, will be covered in the next chapter.

Properties of Simple Harmonic Motion

Consider a particle P moving in a straight line with an acceleration that is directed towards O, a fixed point on the line, the magnitude of the acceleration being proportional to the distance OP.

When the displacement of P from O is x, the acceleration is in the negative direction, i.e. $\ddot{x}$ is negative and we can write $\ddot{x} = -n^2x$. (We use n^2 for the constant of proportion because n^2 cannot be negative.)

Similarly if P is at a point where x is negative, $\ddot{x}$ is in the positive direction so again $\ddot{x} = -n^2x$.

The equation $\ddot{x} = -n^2x$ is the basic equation of SHM.
Any motion that satisfies this equation is known to be simple harmonic.

Further, if the velocity, $\dot{x}$, is zero at a point A, distant a from O, the following diagram shows the basic information about SHM.

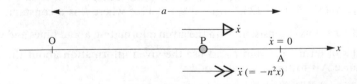

As can be seen from the description of the motion, the symbol a is used for a distance in this topic. This means that we cannot use a to indicate the acceleration so $\ddot{x}$ is usually used instead.

The acceleration is a function of x, so we will use $\ddot{x} = v\dfrac{dv}{dx}$ and because it is directed in the negative sense we have

$$\ddot{x} = -n^2 x \qquad \Rightarrow \qquad v\dfrac{dv}{dx} = -n^2 x \qquad [1]$$

Hence
$$\int v\ dv = -n^2 \int x\ dx$$

$$\Rightarrow \qquad \tfrac{1}{2}v^2 = -\tfrac{1}{2}n^2 x^2 + K$$

$v = 0$ when $x = a$, $\qquad \Rightarrow \qquad K = \tfrac{1}{2}n^2 a^2$

$$\therefore \qquad v^2 = n^2(a^2 - x^2) \qquad [2]$$

From [2] we can see that the greatest value of v is given when $x = 0$

i.e. $\qquad\qquad\qquad v_{max} = na$

Equation [2] also shows that $v = 0$ when $x = \pm a$, so the particle oscillates between two points A and A', on opposite sides of O and each distant a from O.

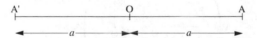

The point O is called the centre, or mean position, of the SHM and the distance OA is called the *amplitude* of the motion.

Examples 8e

1. A particle is describing SHM with amplitude 2 m. If its speed is $3\,\text{m}\,\text{s}^{-1}$ when it is 1 m from the centre of the path find

(a) the basic equation of the SHM being described,

(b) the maximum acceleration,

(c) the speed when the particle is 1.5 m from the centre of the path.

(a) We know that $a = 2$ and $v = 3$ when $x = 1$.

Using $\qquad v^2 = n^2(a^2 - x^2)$ gives

$$9 = n^2(4 - 1) \qquad \Rightarrow \qquad n^2 = 3$$

$\therefore$ the basic equation of the SHM is $\ddot{x} = -3x$

(b)   From $\ddot{x} = -3x$ we see that $\ddot{x}$ is greatest when x is greatest.

The greatest value of x is 2

∴   the maximum acceleration is $6\,\mathrm{m\,s^{-2}}$ towards O.

(c)   When $x = 1.5$, $v^2 = 3(2^2 - 1.5^2) = 5.25$

∴   the speed when $x = 1.5$ is $2.29\,\mathrm{m\,s^{-1}}$ (3 sf)

ASSOCIATED CIRCULAR MOTION

There is a very interesting link between circular motion and SHM which we are now going to look at.

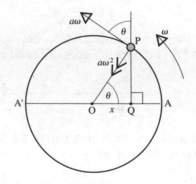

The diagram shows a particle P travelling at a constant angular velocity ω, round a circle with centre O and radius a. Q is the projection of P on the diameter AA′ (i.e. Q is the foot of the perpendicular from P to AA′).

We know from the work on circular motion that the acceleration of P is $a\omega^2$ towards the centre.

Now the component in the direction AA′ of this acceleration gives the acceleration of Q.

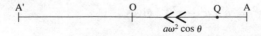

Therefore the acceleration of Q is $a\omega^2 \cos\theta$ in the direction $\overrightarrow{\text{QO}}$

From triangle OPQ, $\cos \theta = \dfrac{x}{a}$

Therefore the acceleration of Q towards O is $a\omega^2 \left(\dfrac{x}{a}\right)$,   i.e. $\omega^2 x$.

But ω^2 is constant so we see that the acceleration of Q is proportional to the distance of Q from O, and is always towards O.

Therefore Q describes SHM about O as centre and with amplitude a.

**As a point P travels round a circle at constant angular speed ω,
its projection on a diameter of the circle describes SHM
with equation $\ddot{x} = -\omega^2 x$**

Comparing this equation of SHM with the standard equation $\ddot{x} = -n^2 x$
we see that ω is equivalent to the constant n, used earlier.

Further properties of SHM can now be discovered by considering the associated circular motion.

● We can find period of the oscillations.

**The time, T, taken to describe one complete oscillation
(from A to A′ and back to A)
is called the *periodic time*, or simply *the period*.**

As Q describes a complete oscillation, P performs one complete revolution of the circle.

The angular velocity of P is ω radians per second.

In one revolution P turns through 2π radians.

So the time taken to describe the revolution is $\dfrac{2\pi}{\omega}$

Therefore T, the period or periodic time of the SHM, is given by $T = \dfrac{2\pi}{\omega}$

But as $\omega = n$,   the period is also given by

$$T = \frac{2\pi}{n}$$

Note that the period is independent of the amplitude of the motion.

● We can find an expression for x in terms of time.

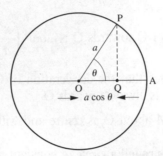

If we measure time from when P is at A, i.e. $t = 0$ when $x = a$, then, t seconds after leaving A, P will have turned through ωt radians, i.e. $\theta = \omega t$.

By this time Q has reached the point where $x = a \cos \theta$.

Therefore, t seconds after leaving the end of the path, P is in a position where $x = a \cos \omega t$. But $\omega = n$ therefore

$$x = a \cos nt$$

Remember that, in this formula, x is measured from O but t is measured from A.

● The expression $x = a \cos nt$ can be used to check the expression derived for the velocity of P on p. 211.

Differentiating x with respect to time gives $\dot{x} = -an \sin nt$

In triangle OPQ, $PQ^2 = a^2 - x^2$

therefore $\sin \omega t = \dfrac{\pm\sqrt{(a^2 - x^2)}}{a}$ $\Rightarrow$ $\dot{x} = v = \pm n\sqrt{(a^2 - x^2)}$

∴ $$v^2 = n^2(a^2 - x^2)$$

Note that v can be either positive or negative. This is because P passes through any particular point in both the positive and the negative direction.

Although it is not necessary for the reader to memorise the details of the background results and their derivations given in this chapter, it is very useful to understand how information about SHM can be obtained by looking at the simple case of motion in a circle with constant angular velocity. This approach often provides a simple solution to a problem.

In any case the basic properties of SHM should be known and can be quoted.

Summary of Definitions and Relationships for SHM

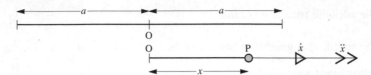

AA' is the path

O is the centre, or mean position

a is the amplitude

An oscillation is the journey from A to A' and back to A

The period of an oscillation is T where $T = \dfrac{2\pi}{n}$

For a general position of the particle P

$$\ddot{x} = -n^2 x \quad \text{where } n \text{ is a constant}$$
$$\dot{x} = n^2(a^2 - x^2)$$
$$x = a \cos nt \quad \text{where} \quad t = 0 \quad \text{when} \quad x = a$$

The maximum acceleration occurs at A and A' and its magnitude is $n^2 a$.

The maximum speed is na, occurring at O.

Examples 8e (continued)

2. A particle P is describing SHM of amplitude 2.5 m. When P is 2 m from the centre of the path, the speed is $3\,\text{m s}^{-1}$. Find

(a) the periodic time of the oscillations

(b) the greatest speed

(c) the magnitude of the greatest acceleration.

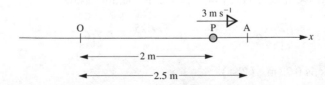

First we will find the value of n as this occurs in every formula for SHM.

Using $v^2 = n^2(a^2 - x^2)$ with $a = 2.5$ and $v = 3$ when $x = 2$

gives $\quad 3^2 = n^2(2.5^2 - 2^2) \qquad \Rightarrow \qquad n^2 = \dfrac{9}{2.25}$

$\therefore \qquad n = 2$

(a) $T = \dfrac{2\pi}{n}$ $\Rightarrow$ $T = \pi$

The periodic time is π seconds.

(b) The speed, v, is greatest when $x = 0$,   i.e. $v_{\max} = na$

The greatest speed is $5\,\mathrm{m\,s^{-1}}$.

(c) $\ddot{x} = -n^2 x$

The acceleration is greatest when $x = a$

$\therefore$ the magnitude of the greatest acceleration is $2^2 \times 2.5$,   i.e. $10\,\mathrm{m\,s^{-2}}$.

3. **A piston is performing SHM at the rate of 4 oscillations per minute. The maximum speed of the piston is $0.3\,\mathrm{m\,s^{-1}}$. By modelling the piston as a particle, find the amplitude of the motion. Find also the speed and the acceleration of the piston when it is $0.5\,\mathrm{m}$ from the centre of the path.**

Give answers corrected to 2 significant figures.

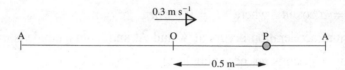

There are 4 oscillations per minute so 1 oscillation takes 15 seconds

$\therefore$ $T = 15$ $\Rightarrow$ $\dfrac{2\pi}{n} = 15$ $\Rightarrow$ $n = \dfrac{2\pi}{15} = 0.4188\ldots$

In this type of solution where the value of n is often used in subsequent calculations, you can, as an alternative to storing it in your calculator or retaining 4 significant figures, use $\frac{2\pi}{15}$ whenever n occurs.

The maximum speed is given by na where a is the amplitude.

Therefore $0.3 = \left(\dfrac{2\pi}{15}\right)a$ $\Rightarrow$ $a = \dfrac{2.25}{\pi} = 0.7161\ldots$

The amplitude is $0.72\,\mathrm{m}$ (2 sf)

The speed $v\,\mathrm{m\,s^{-1}}$ is given by $v = n\sqrt{(a^2 - x^2)}$

When $x = 0.5$, $v = \dfrac{2\pi}{15}\sqrt{\left(\left[\dfrac{2.25}{\pi}\right]^2 - 0.5^2\right)} = 0.214\ldots$

The required speed is $0.21\,\mathrm{m\,s^{-1}}$ (2 sf)

When $x = 0.5$, the acceleration is $n^2 x$, i.e. $\left(\dfrac{2\pi}{15}\right)^2 \times 0.5$

The required acceleration is 0.088 m s^{-2} (2 sf)

4. **A, B and C, in that order, are three points on a straight line and a particle P is moving on that line with SHM. The velocities of P at A, B and C are zero, 2 m s^{-1} and -1 m s^{-1} respectively. If AB $= 1 \text{ m}$ and AC $= 4 \text{ m}$, find the amplitude of the motion and the periodic time.**

The velocity at A is zero so A must be at one end of the path (we will take it as the left-hand end). The signs of the velocities at B and C show that P is moving away from A when at B and back towards A when at C. *Speed* at B $>$ *speed* at C, so B is nearer to the centre than C is. The length of the path is greater than 4 cm, so the amplitude is greater than 2 cm. From all these facts we see that B and C are on opposite sides of the centre O.

We will take a metres as the amplitude and measure x from O.

$v = 2$ when $x = -(a - 1)$

$v = -1$ when $x = 4 - a$

Using $v^2 = n^2(a^2 - x^2)$ gives

$$4 = n^2[a^2 - (1 - a)^2] \quad \Rightarrow \quad 4 = n^2[2a - 1] \qquad [1]$$

and $1 = n^2[a^2 - (4 - a)^2] \quad \Rightarrow \quad 1 = n^2[8a - 16]$ [2]

$[1] - 4 \times [2]$ gives

$$n^2[2a - 1] - 4n^2[8a - 16] = 0 \quad \Rightarrow \quad 30a = 63$$

$$\Rightarrow \quad a = 2.1$$

The amplitude is 2.1 m

From [1], $4 = n^2(4.2 - 1) \quad \Rightarrow \quad n = 1.118\ldots$

The period, T, is given by $\dfrac{2\pi}{n}$

$\therefore \qquad T = \dfrac{2\pi}{1.118\ldots} = 5.619\ldots$

The periodic time is 5.62 seconds   (3 sf)

5. A particle P is moving with SHM of period 3π seconds, on a path with centre O and amplitude $6\,\text{m}$. The particle starts from A, one end of the path.

(a) Find the time taken for P to travel from A to B, a distance of $3\,\text{m}$.

(b) Find the distance AC if P travels from A to C in $\frac{2}{3}\pi$ seconds.

(a) 1st Method – using associated circular motion.

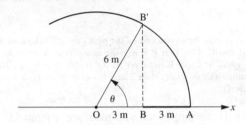

SHM from A to B corresponds to circular motion from A to B'.

When P is at B, $\cos\theta = \frac{3}{6} = \frac{1}{2}$

$\Rightarrow \quad \theta = \frac{1}{3}\pi$ which is $\frac{1}{6}$ of a revolution.

The periodic time is equal to the time it takes B' to complete 1 revolution.

$\therefore$ the time taken to travel arc AB' is $\frac{1}{6}$ of the periodic time,

i.e. $\frac{1}{6} \times 3\pi$ seconds

The time it takes P to travel along AB with SHM is $\frac{1}{2}\pi$ seconds

2nd Method – using the formula $x = a \cos nt$

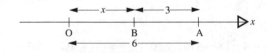

$x = \text{OB} = 6 - 3 = 3$

$$T = \frac{2\pi}{n} \quad \Rightarrow \quad n = \frac{2\pi}{3\pi} \quad \Rightarrow \quad n = \frac{2}{3}$$

$x = a \cos nt$ gives $3 = 6 \cos \dfrac{2t}{3}$

$\therefore \quad \cos\dfrac{2t}{3} = \dfrac{1}{2} \quad \Rightarrow \quad \dfrac{2t}{3} = \dfrac{\pi}{3} \quad \Rightarrow \quad t = \dfrac{\pi}{2}$

$\therefore$ time taken to travel AB with SHM is $\frac{1}{2}\pi$ seconds.

(b) 1st Method

Time to travel round the circumference $= 3\pi$ s.

Time taken to travel round arc $AC' = \frac{2}{3}\pi$ s

$\qquad\qquad\qquad\qquad\qquad = \frac{2}{9}$ of 3π s

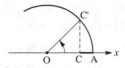

$\therefore$ arc $AC' = \frac{2}{9}$ of the circumference

$\therefore$ $A\hat{O}C' = \frac{2}{9}$ of 2π rad $= 4\pi/9$ rad

$OC = 6 \cos AOC' = 1.0418\ldots$ $\Rightarrow$ $AC = 6 - OC = 4.958\ldots$

$\therefore$ the distance AC is 4.96 m (3 sf)

2nd Method

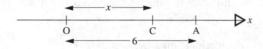

As found in part (a), $n = \frac{2}{3}$

Using $x = a \cos nt$ when $n = \frac{2}{3}$ and $t = \frac{2}{3}\pi$

gives $x = 6 \cos\left(\frac{2}{3}\right)\left(\frac{2}{3}\pi\right) = 6 \cos \frac{4}{9}\pi$

$\Rightarrow \qquad x = 1.0418\ldots$

But x m is the distance OC

$\therefore$ the distance AC is 4.96 m (3 sf)

EXERCISE 8e

In questions 1 to 8 a particle P is performing simple harmonic motion in a straight line as shown. All units are consistent and based on metres and seconds.

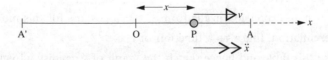

1. When $\ddot{x} = -9x$ and the amplitude is 5 m, find

(a) the period

(b) the maximum speed

(c) the speed when $x = -2$.

2. If the maximum acceleration is $10\,\mathrm{m\,s^{-2}}$ and the maximum speed is $8\,\mathrm{m\,s^{-1}}$, find

   (a)   the period

   (b)   the amplitude

   (c)   the speed when $x = 4$

   (d)   the acceleration when $x = 4$.

3. Given that $v = 4.8$ when $x = 0.7$ and $v = 3$ when $x = 2$, find

   (a)   the amplitude

   (b)   the period

   (c)   the maximum speed

   (d)   the maximum acceleration.

4. If the period is 5π seconds and the maximum speed is $2\,\mathrm{m\,s^{-1}}$, find

   (a)   the amplitude

   (b)   the speed and the acceleration when $x = -1$.

5. Given that the frequency of oscillation is 0.25 oscillations per second, and that $v = 0.5$ when $x = 0.2$ find

   (a)   the period

   (b)   the amplitude

   (c)   the speed and the acceleration when $x = 0.15$.

6. The motion is defined by $x = 4\cos 3t$.

   (a)   Find $\dot{x}$.

   (b)   Find $\ddot{x}$.

   (c)   Use your result from (b) to express $\ddot{x}$ in terms of x.

   (d)   What is the period of the motion?

7. Given that $\ddot{x} = -\left(\dfrac{\pi^2}{16}\right)x$ and the amplitude is $0.5\,\mathrm{m}$,

   (a)   write down an expression in the form $x = a\cos nt$ for this motion,

   (b)   by differentiation, find v as a function of t,

   (c)   complete this table and hence sketch the graph of v against x for the range $1 \geqslant x \geqslant -1$.

t	0	1	2	3	4	5	6	7	8
x									
v									

8. When the period of the motion is 8 seconds, use $x = a \cos nt$ to find the time taken to go from the point where $x = a$ to the point where $x = \frac{1}{2}a$.

9. A particle performs two SHM oscillations each second. Its speed when it is 0.02 m from its mean position is half the maximum speed. Find the amplitude of the motion, the maximum acceleration and the speed at a distance 0.01 m from the mean position.

The solutions to questions 10 to 12 may be based either on the standard formulae for SHM or on the use of the associated circular motion.

10. A particle is travelling between two points P and Q with simple harmonic motion. If the distance PQ is 6 m and the maximum acceleration of the particle is 16 m s^{-2}, find the time taken to travel

   (a)   a distance 1.5 m from P

   (b)   from P to the midpoint O of PQ

   (c)   from the midpoint of PO to the midpoint of OQ.

11. A particle describes simple harmonic motion between two points A and B. The period of one oscillation is 12 seconds. The particle starts from A and after 2 seconds has reached a point distant 0.5 m from A. Find

   (a)   the amplitude of the motion

   (b)   the maximum acceleration

   (c)   the velocity 4 seconds after leaving A.

12. A particle is performing simple harmonic motion of amplitude 0.8 m about a fixed point O. A and B are two points on the path of the particle such that $OA = 0.6$ m and $OB = 0.4$ m. If the particle takes 2 seconds to travel from A to B find, correct to one decimal place, the periodic time of the SHM if

   (a)   A and B are on the same side of O

   (b)   A and B are on opposite sides of O.

13. The prongs of a tuning fork, which sounds middle C, are vibrating at a rate of 256 oscillations per second. Assuming that the motion of the prongs is simple harmonic and that the amplitude of the end of a prong is 0.1 mm, find

   (a)   the maximum velocity and the maximum acceleration of the end of a prong

   (b)   the velocity and acceleration of the end of a prong when its displacement from the centre of its path is 0.05 mm.

Mixed Problems

The questions in this exercise involve a variety of methods and ideas.

EXERCISE 8f

1. A cylindrical buoy, of length l metres, is held at rest, with its lower plane face touching the surface of the water and is then released. After t seconds its lower plane face is at a depth x metres below the surface, its velocity is $v\,\mathrm{m\,s^{-1}}$ and its acceleration is $a\,\mathrm{m\,s^{-2}}$. Because the upward force exerted by the water is known to be proportional to x, the motion may be modelled by the equation

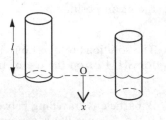

$$ a = g\left(1 - \frac{k}{l}x\right), \quad \text{where, for this particular buoy,} \quad k = 3. $$

   (a) Find v^2 in terms of x.

   (b) Find the value of x when $v = 0$.

   (c) This method of placing the buoy in the water is unsatisfactory as it springs back up. If it is to be placed gently at a lower depth where it will remain in equilibrium, what is the value of x in this position?

2. A body is fired vertically into space from the surface of the earth. After t seconds its displacement from the centre of the earth is s metres and its velocity is v metres per second. Its acceleration $a\,\mathrm{m\,s^{-2}}$ is given by

$$ a = -\frac{k}{s^2} \quad \text{where } k \text{ is a constant.} $$

   Initially $s = r$ and $v = u$.

   (a) Show that

$$ \frac{1}{2}(v^2 - u^2) = k\left(\frac{1}{s} - \frac{1}{r}\right) $$

   (b) Given that $k = 4 \times 10^{14}$ and $r = 6.4 \times 10^6$, find the change in kinetic energy of a satellite of mass 2000 kg in going from the surface of the earth to a point where $s = 6.8 \times 10^6$.

3. It is thought that the motion of a man running a 100 m race may be modelled by the equation $a = \lambda - \mu t$ where $a\,\mathrm{m\,s^{-2}}$ is his acceleration after t seconds from the start and λ and μ are constants.

   (a) Find expressions for his speed and the distance covered at time t.

   (b) Suggest a practical way in which values for all the unknown constants (λ, μ and any others) may be found.

   (c) Suggest a way in which the model could then be assessed for validity.

4. A student is considering a model for the motion of a train on a light railway between two stopping points A and B. He has the following data.

1. The journey time from A to B is 1200 seconds.
2. The train accelerates from rest for 60 seconds to a speed of $30\,\text{m}\,\text{s}^{-1}$ and then decelerates for 60 seconds, coming to rest at B.
3. The distance between A and B is 1900 metres.

On the basis of this data he constructs this model.

First model   The acceleration is constant and the deceleration is also constant.

Another piece of data is then obtained,

4. 30 seconds after leaving A the train has a speed of $17\,\text{m}\,\text{s}^{-1}$ and has travelled 200 metres.

As a result the student constructs a different model.

Second model   The displacement of the train from A after t seconds is s metres and the motion is modelled by the equation

$$k\frac{\mathrm{d}s}{\mathrm{d}t} = t^2(120 - t)^2, \quad \text{where } k \text{ is a constant.}$$

(a) Verify that the second model gives zero acceleration when $t = 60$.
(b) By considering the maximum velocity ($30\,\text{m}\,\text{s}^{-1}$), find the value of k.
(c) Find the values given for the distance AB
  (i) by the first model
  (ii) by the second model.
(d) Find the values given for the velocity after $30\,\text{s}$
  (i) by the first model
  (ii) by the second model.
(e) Find the values given for the distance travelled after $30\,\text{s}$
  (i) by the first model
  (ii) by the second model.
(f) Do you think that either of these models fits the data reasonably well or would you look for a better model?

5. On a particular day in a harbour, high tide at its entrance occurs at noon and the water depth is then 11 m. Low tide occurs $6\frac{1}{4}$ hours later and the water depth is then 5 m. The water level performs simple harmonic motion.

(a) Find the amplitude and period of this motion.
(b) Find the time when the water level will be falling at its maximum rate and find this rate in metres per minute.
(c) A ship needs a depth of 7 m to enter the harbour. Find the latest time after noon at which it can enter without having to wait for low tide to pass.
(d) If it arrives after this time, what is the next time when it can enter harbour?

6. An astronomer observes a small body which moves towards the planet Jupiter and eventually passes in front of it. This body appears to be moving in simple harmonic motion along the straight line AB. He thinks that it is a moon orbiting the planet and that he is viewing it along the plane of the orbit. He knows that the radius of the planet is 71 600 km.

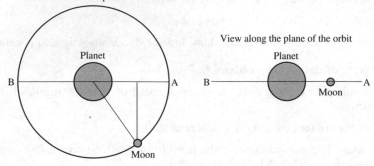

Waiting until it reaches a position in front of point C, which is the centre of the planet, he then records the time it takes to travel from there to two other positions. These are

  (i)   in front of point D, which is at the edge of the planet,

 (ii) at B, just before it appears to return along the line BA.

Points used in observations

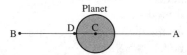

His results are in the table, where t is the time in minutes from when the moon appeared in front of C.

Position	C	D	B
t	0	69	637

(a) Find the period in minutes.

(b) Find the angular velocity of the moon about the planet in radians per minute.

(c) Find the radius of the orbit.

     (This is also the amplitude of the apparent oscillation.)

CHAPTER 9

VARIABLE FORCE

THE RELATIONSHIP BETWEEN FORCE AND ACCELERATION

Newton's Law of Motion applies to *any* motion, however it is caused. It applies whether the force producing the motion is constant or variable, i.e.
for a body of mass m, moving under the action of *any* force F, then $F = ma$.

When F varies in a specified way, it may be possible to represent the motion by a differential equation.

If F is a function of time, i.e. $F = f(t)$, then using $a = \dfrac{dv}{dt}$ gives

$$f(t) = m\frac{dv}{dt}$$

$\Rightarrow$
$$\int f(t)\ dt = \int m\ dv$$

If F is a function of displacement, i.e. $F = f(s)$, then using $a = v\dfrac{dv}{ds}$ gives

$$f(s) = mv\frac{dv}{ds}$$

$\Rightarrow$
$$\int f(s)\ ds = \int mv\ dv$$

The Work Done by a Variable Force

Suppose that a particle moving under the action of a force F, where $F = f(s)$, has a velocity u when $s = 0$ and a velocity v after covering a distance s.

The relationship $F = mv\dfrac{dv}{ds}$ becomes $\displaystyle\int_0^s F\ ds = \int_u^v mv\ dv$

giving
$$\int_0^s F\ ds = \tfrac{1}{2}mv^2 - \tfrac{1}{2}mu^2$$

225

Now $\frac{1}{2}mv^2 - \frac{1}{2}mu^2$ is the increase in KE of the particle and this is equal to the amount of work done in causing it.

Therefore $\int F\,ds$ represents the work done by the force F, i.e.

> **when a variable force F, where $F = f(s)$, moves its point of application through a distance s, the work done by the force is given by $\int f(s)\,ds$**

Note that, when F is constant, this result gives Work done $= Fs$ which was derived in Module E.

The Impulse Exerted by a Variable Force

Again we consider a force F that causes the velocity of a particle to increase from u to v, but this time the increase takes place over an interval of time t.

Using the equation $F = m\dfrac{dv}{dt}$ gives $\displaystyle\int_0^t F\,dt = \int_u^v m\,dv$

giving $\displaystyle\int_0^t F\,dt = mv - mu$

But $mv - mu$ is the increase in momentum of the particle over the time t and we know that this is equal to the impulse of the force producing it.

Therefore $\int F\,dt$ represents the impulse of the force F, i.e.

> **when a variable force F, where $F = f(t)$, acts on an object for a time t the impulse exerted by the force is given by $\int f(t)\,dt$**

When F is constant, this gives Impulse $= Ft$, as used in Module E.

Examples 9a

1. **A particle P is moving in a straight line under the action of a variable force F. The particle passes through a point O on the line and t seconds later its displacement s from O is given by $s = r\sin 4t$. Show that the force is proportional to the displacement of P from O and describe the behaviour of the force.**

$$s = r \sin 4t$$

$$\therefore \quad v = \frac{ds}{dt} = 4r \cos 4t$$

$$\therefore \quad a = \frac{dv}{dt} = -16r \sin 4t$$

Newton's Law gives $F = m(-16r \sin 4t)$ $\Rightarrow$ $F = (-16m)s$

As $16m$ is constant, F is proportional to s

The relationship between F and s can be expressed as $F = 16m(-s)$. In this form, as $16m$ is positive, we see that F and s are of opposite sign.

Therefore as P moves away from O in either direction, the force F acts towards O and is proportional to the distance of P from O.

2. **A particle P, of mass 1 kg, is moving horizontally along the x-axis and passes through the origin O with speed $2\,\mathrm{m\,s}^{-1}$. A force F acts on P in the positive direction and F is equal to $2x$. When P has moved through $3\,\mathrm{m}$, find**

   (a) **the work done by F**   **(b)** **the speed of P.**

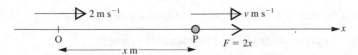

(a)   The work done by F is given by $\int F\,dx$, where $F = 2x$.

$$\therefore \quad \text{when } x = 3, \text{ work done} = \int_0^3 2x\,dx$$

$$= \left[x^2 \right]_0^3$$

The work done by F is $9\,\mathrm{J}$

(b)   P is moving horizontally so there is no change in its PE

Therefore using   Work done = Increase in ME   gives

$$9 = \tfrac{1}{2}mv^2 - \tfrac{1}{2}mu^2$$

$$= \tfrac{1}{2}v^2 - \tfrac{1}{2}(4)$$

$$\Rightarrow \quad v^2 = 22$$

The speed of P is $4.69\,\mathrm{m\,s}^{-1}$ (3 sf)

3. **A body of mass 1.4 kg falls from rest in a medium which exerts a resistance of $(2t + k)\,\text{N},$ where k is a constant. The speed of the body after falling for 4 seconds is $18\,\text{m s}^{-1}.$ Find**

   (a)   the value of k

   (b)   the speed after a further 3 seconds.

The resultant force F acting downwards on the body is $mg - R$

i.e. $1.4 \times 9.8 - (2t + k)$

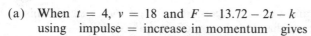

(a)   When $t = 4$, $v = 18$ and $F = 13.72 - 2t - k$
      using   impulse $=$ increase in momentum   gives

$$\int_0^4 (13.72 - 2t - k)\,\mathrm{d}t = 1.4(18 - 0)$$

$$\therefore \qquad \left[13.72t - t^2 - kt \right]_0^4 = 25.2$$

$$\Rightarrow \qquad\qquad\qquad 4k = 13.68$$

$$\therefore \qquad\qquad\qquad k = 3.42$$

(b)   After a further 3 seconds, $t = 7$ and the velocity is $v\,\text{m s}^{-1}.$

   Using   impulse $=$ increase in momentum   gives

$$\therefore \qquad \left[13.72t - t^2 - 3.42t \right]_0^7 = 1.4v$$

$$\Rightarrow \qquad\qquad\qquad v = 16.5$$

The speed after 7 seconds is $16.5\,\text{m s}^{-1}$

4. **A bag of ballast of mass m is dropped from a stationary hot-air balloon.**

   (a)   By modelling the ballast as a particle and the air as a medium whose resistance to motion is $0.1v$, where v is the velocity of the ballast, find an expression for the velocity of the ballast after an interval of time t.

   (b)   Comment on the assumptions that have been made in choosing this model.

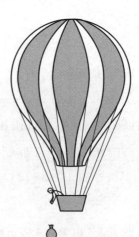

(a) The resultant downward force acting on the ballast at any time t is $mg - 0.1v$

Newton's law $F = ma$ gives

$$mg - 0.1v = ma$$

Using $a = \dfrac{dv}{dt}$ gives $mg - 0.1v = m\dfrac{dv}{dt}$

$\therefore$

$$m \int \frac{1}{mg - 0.1v}\,dv = \int dt$$

$\therefore$

$$-\frac{m}{0.1}\ln A(mg - 0.1v) = t \quad \text{(using ln } A \text{ as the constant of integration)}$$

$\Rightarrow$

$$-m \ln A(mg - 0.1v) = 0.1t$$

When $t = 0$, $v = 0$, therefore $-m \ln Amg = 0$

$\therefore \quad Amg = 1 \quad \Rightarrow \quad A = \dfrac{1}{mg}$

Hence $-m \ln \left(\dfrac{mg - 0.1v}{mg} \right) = 0.1t$

$\therefore \quad \ln \left(\dfrac{mg - 0.1v}{mg} \right) = \dfrac{-t}{10m} \quad \Rightarrow \quad mg - 0.1v = mg\,e^{-t/10m}$

$\therefore \quad v = 10mg(1 - e^{-t/10m})$

(b) It is assumed that the resistance to motion is the same irrespective of the size and shape of the object moving through it. In addition it is assumed that the resistance is related to the velocity in the same way at all times. Both of these assumptions could be unreliable as the bag would probably turn and twist presenting different surfaces to the air below it.

EXERCISE 9a

In questions 1 to 5 a particle P, of mass m, is moving along a straight line and O is a fixed point on that line. At a time t the force acting on P is F, the acceleration of P is a, the velocity is v and the displacement from O is s. All units are consistent.

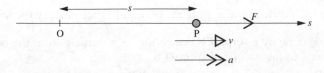

1. If $v = 15t - e^{3t}$ and $m = 2$, find

(a) F in terms of t

(b) the value of F when $t = 0$.

2. Given that $s = \dfrac{1}{t+1}$ and $m = 4$, find

   (a) v in terms of t (b) F in terms of t.

3. The force F is given by $F = 24 - 6e^{-2t}$, and $v = 5$ when $t = 0$. If $m = 3$ find

   (a) a in terms of t (b) v in terms of t.

4. Given that $m = 3$ and that $v = s + \dfrac{1}{s}$ for $s > 0$,

   (a) find F in terms of s

   (b) find a positive constant b such that $F = 0$ when $s = b$

   (c) show that F opposes the motion when $0 < s < b$, and is in the direction of the motion when $s > b$.

5. The force F newtons produces power of 120 W. The mass m is 3 kg and, when $t = 2$, $v = 14$ and $s = 20$.

   $$\left(\text{Power is the rate of doing work, i.e. } F\,\frac{ds}{dt} \text{ or } Fv. \right)$$

   (a) Express F in terms of v.   (c) Find v in terms of s.

   (b) Find v in terms of t.   (d) Find the values of v and s when $t = 8$.

6. A particle P, of mass 5 kg is moving along a straight line and O is a fixed point on that line. Initially P is at rest at a point A on the line where $OA = 2\,m$. After t seconds the displacement of P from O is x metres and its velocity is $v\,\text{m\,s}^{-1}$. A force F acts on P, where $F = -90 \cos 3t$.

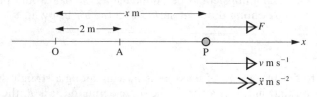

   (a) Find v in terms of t.

   (b) Find x in terms of t.

   (c) Express the acceleration, $\ddot{x}$, in terms of x and hence identify the motion of P.

7. Find the work done by a force $F\,$N which moves a particle from the origin O along the x-axis to the point 5 m from O if

   (a) F is of constant magnitude 10,   (b) $F = x^2$.

8. A body of mass 4 kg falls from rest for 2 seconds in a medium whose resistance is $(2 + 15t)\,$N. Find the velocity at the end of this time. [Take g as 10]

9. A body of mass m falls from rest against a resistive force equal to $\frac{1}{10}s$ where s is the distance fallen. Find the work done by the resistance when the body has fallen a distance d. What is the kinetic energy of the body in this position?

10. A force acts on a particle of mass m causing its speed to increase from u to $2u$ in T seconds. The force acts in a straight line in the direction of motion and is of magnitude kt, where $t = 0$ when the speed is u. Find T in terms of m, u and k.

11. A particle P of mass m passes through the origin O with speed V and moves along the positive x-axis. It is subjected to a retarding force R which causes an acceleration of magnitude kx towards O. When P reaches the point where $x = 4$ find:

   (a) the work done by R

   (b) the speed of P, giving your answers in terms of k and V.

12. A parachutist, of total mass 75 kg, descends vertically in free fall for a period before opening his parachute. Take his initial velocity as zero and model the air resistance by the expression $0.5v^2$.

   (a) Find the greatest speed he can reach in free fall.

   (b) Find s in terms of v.

   (c) Find the distance he has fallen when his speed is $30\,\mathrm{m\,s^{-1}}$.

FORMING, TESTING AND IMPROVING MODELS

A model is formed by making assumptions in order to use a known physical law. The formula can then be used to predict results which, in some cases, can be checked against observed results.

Testing a Model

Suppose that we wish to test the reliability of the model used in Example 9a/4.

The model might be tested as follows.

The equation derived from the model, for v as a function of t, can be integrated to give s in terms of t, i.e. $s = f(t)$.

The time taken for the ballast to hit the ground can be measured and used in $s = f(t)$ to predict the distance fallen. This predicted value can be compared with the actual distance fallen (taken from the altimeter of the balloon). If they are reasonably close we would be encouraged to think that the model is fairly satisfactory but it would be more conclusive if the test were carried out at different heights. Further measurements may confirm the reliability of the model; on the other hand we might find that the correlation between predicted and observed values is not good enough.

Improving a Model

If the predicted and observed results are not close enough for the accuracy required, the next step is identify the feature(s) in the model that may be the cause of the discrepancy.

In this example we have already commented that the assumption about the resistance is suspect and we could try representing the resistance by a different function. We might, for example, increase the constant 0.1, or we could consider taking the resistance as $0.1v^2$.

Whether or not the model has been improved by the adjustment can only be determined by testing the new model and comparing its predictions against observed values.

The process of testing and refining a model can be shown by a flowchart.

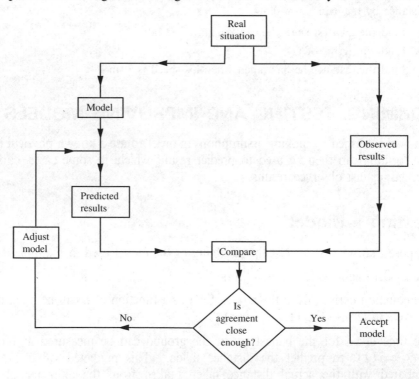

Sometimes when a mathematical model of a situation is constructed, the results it gives appear to correlate well with observed results over the range of circumstances within which it can be tested. If, at a later stage, that range is widened, the model may not be so reliable in the extended conditions and an adjustment may then be necessary.

Consider the following example.

For a project a group of students investigated the vertical motion of a ball of mass 1 kg. They borrowed a ball-projecting machine from the local cricket club, which can give the ball known initial speeds. For each speed the students measured the time taken for the ball to reach its highest point and the following table gives the values recorded, corrected to 2 sf.

Initial speed $V \mathrm{m\,s^{-1}}$	8	12	15	21	26	35
Time, Ts, to highest point	0.8	1.1	1.2	1.7	2.1	2.7

They took $10 \mathrm{m\,s^{-2}}$ as the acceleration and their first model assumed zero air resistance.

Using $v = u + at$ they calculated the value of T given by this model for each value of V and entered these results in the table for comparison with the observed results, inserting these values of T in a third row.

Initial speed $V \mathrm{m\,s^{-1}}$	8	12	15	21	26	35
Time, Ts, to highest point	0.8	1.1	1.2	1.6	2.0	2.7
Time given by first model	0.8	1.2	1.5	2.1	2.6	3.5

The students were not satisfied with this correlation because, although there was good agreement between the calculated and observed values for the lower values of V, the correlation decreased as V increased, so they tried another model.

The second model assumed that air resistance is given by kv where k is a constant.

Using $F = ma$ and $a = \dfrac{dv}{dt}$ gives

$$-(g + kv) = \frac{1 dv}{dt} \quad \Rightarrow \quad -\int \frac{1}{g + kv} \, dv = \int dt$$

$\therefore$
$$-\left[\frac{1}{k} \ln (g + kv) \right]_V^0 = \left[t \right]_0^T$$

$\therefore$
$$T = \frac{-1}{k} [\ln g - \ln (g + kV)] = \frac{1}{k} \ln \left(\frac{g + kV}{g} \right)$$

Using the observed value of T when $V = 21$, the students found, by trial and improvement, that 0.21 is a reasonable value for k. The resulting values of T were compared with the observed results as before.

Initial speed $v \mathrm{m\,s^{-1}}$	8	12	15	21	26	35
Time, Ts, to highest point	0.8	1.1	1.2	1.7	2.1	2.7
Time given by second model	0.7	1.1	1.3	1.7	2.1	2.6

The students were satisfied with the agreement between the observed figures and those predicted by the second model, for the range of values of V tested.

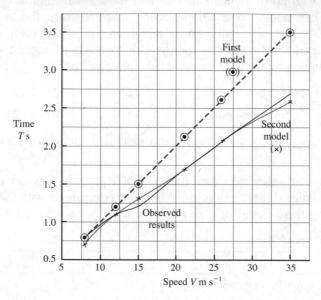

EXERCISE 9b

1. The velocity–time graph illustrates the motion of a car, of mass 800 kg, whose velocity is $v\,\mathrm{m\,s^{-1}}$ at time t seconds. The car starts from rest, accelerates to a speed of $15\,\mathrm{m\,s^{-1}}$ in 20 s and then continues at this speed. Its acceleration decreases as the speed of $15\,\mathrm{m\,s^{-1}}$ is approached. A mathematical model is sought that will give the value of the resultant force, F newtons, at various times. A suitable model is thought to be $F = p - qt$ and the values of p and q that fit the given data are required.

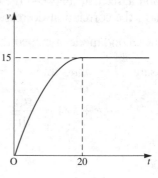

(a) Using the fact that the acceleration approaches zero as $t \to 20$, find a relationship between p and q.

(b) By integrating $F = p - qt$, find v in terms of p, q and t.

(c) Using the result of part (b), together with the data given on velocity and time, find another relationship between p and q.

(d) Find the values of p and q using the results of parts (a) and (c).

(e) Find the value predicted by this model for the velocity 10 s after the start and comment on how well this fits the given data.

(f) Do you consider that this model would be appropriate for predicting the value of F when the car accelerates from rest to $30\,\mathrm{m\,s^{-1}}$ in 40 seconds?

2. A cyclist is travelling along a horizontal road exerting a constant forward force of 80 N. The total mass including the bicycle is 80 kg. His velocity at t seconds after he started is v m s^{-1}. He starts from rest when $t = 0$. The total resistance to his motion is to be modelled as proportional to v.

   (a) Given that the maximum speed he reaches is 8 m s^{-1}, find the resistance in terms of v.

   (b) Find v in terms of t.

   (c) Find his velocity after 5 s.

   (d) How long does it take him to reach a speed of
       (i) 7 m s^{-1} (ii) 7.9 m s^{-1} (iii) 8 m s^{-1}

   (e) Using the result of part (b), find s in terms of t and hence find the distance the cyclist travels in the first 5 seconds.

   (f) State, with reasons, whether you think that the model is reliable.

3. Aircraft windscreens were being tested to determine the maximum impulse they could withstand when in collision, in flight, with small objects such as birds. A machine fired objects, of known mass and velocity, at the windscreen and the impulse was measured each time by sensors attached to the screen.

   Suspecting that the sensors might not be accurate, a 'real-life' test was carried out using oven-ready chickens, of standard mass 3 kg, as the missiles. The results given by the sensors were then compared with the values of J calculated by using the model $J = mv$.

Speed v (m s^{-1})	20	40	80	100	150	200
J (Ns) from sensor	67	135	273	342	509	681
J (Ns) from model	60	120	240	300	450	600

   The results did not correlate well so the test was carried out with new sensors. The results from the second set of sensors, however, were very close to those given by the first sensors, so the suitability of the model came under scrutiny instead.

   Careful observation of what actually happened at impact revealed that the 'birds' bounced off the screen. The model was then adapted to take this into account, by estimating the coefficient of restitution as 0.1.

   (a) Form a new model for J, allowing or the bounce.

   (b) Use this model to calculate J for the speeds given in the table (still taking $m = 3$).

   (c) Comment on the correlation between the values predicted by the new model and those given by the sensors.

   (d) Can you suggest a further refinement to the model that might give closer correlation?

4. A material has the property that, when a bullet is fired into it, the resistance to the motion of the bullet increases as the bullet penetrates further. It is thought that this resistance, R newtons, can be modelled by either $R = kx$ or $R = kx^2$, where x is the depth of penetration and k is a constant. Tests are carried out with bullets of mass $0.02\,$kg which are fired horizontally into a fixed block of the material. It is found that a bullet entering at $400\,\mathrm{m\,s^{-1}}$ penetrates to a depth of $0.1\,$m and a bullet entering at $800\,\mathrm{m\,s^{-1}}$ penetrates $0.16\,$m. These test data allow two values for k to be calculated for each model. By doing this, or otherwise, decide which model fits the data more closely. Give an estimate of the value of k for that model.

*5. The acceleration from rest of a car on a level road is being tested. The car is of mass $800\,$kg and its engine is working at a constant rate of $40\,$kW. It is found that after 5 seconds the speed is $20\,\mathrm{m\,s^{-1}}$, and the speed reached $30\,\mathrm{m\,s^{-1}}$ after 12 seconds.

In order to predict the time, t seconds, taken to reach various speeds, $v\,\mathrm{m\,s^{-1}}$, a model is formed assuming a constant resistance, of $800\,$N, to the motion of the car.

(a) (i) Find the maximum speed the car can reach.

   (ii) Show that $\displaystyle\int \frac{v}{50 - v}\, dv = \int dt$

   (iii) Express t in terms of v. $\left(\text{Write } \dfrac{v}{50 - v} \text{ as } \dfrac{(v - 50) + 50}{50 - v}\right)$

   (iv) Find the times predicted by this model for the car to reach $20\,\mathrm{m\,s^{-1}}$ and $30\,\mathrm{m\,s^{-1}}$.

A second model is considered, taking the resistance as kv where k is constant.

(b) (i) Using the maximum speed found in part (i) above, find the value of k.
   (ii) Express t in terms of v.
   (iii) Find the times predicted by this model for the car to reach $20\,\mathrm{m\,s^{-1}}$ and $30\,\mathrm{m\,s^{-1}}$.

(c) By comparing the results predicted by each model with the measured results, comment on the suitability of each model.

A further experimental measurement gives the time taken to reach $35\,\mathrm{m\,s^{-1}}$ as 19 seconds.

(d) (i) Calculate the times predicted to reach $35\,\mathrm{m\,s^{-1}}$.
   (ii) How does this result affect your answer to part (c).

THE LAW OF UNIVERSAL GRAVITATION

Yet another law formulated by Newton concerns the forces of attraction that arise between any two bodies in the universe. It states that if the centres of the bodies are distant r apart and their masses are m_1 and m_2, the force F that attracts each to the other is given by

$$F = \frac{Gm_1 m_2}{r^2}$$

The value of the constant G is approximately 6.7×10^{-11} and rearranging the formula as $G = \dfrac{Fr^2}{m_1 m_2}$ shows that G is measured in $m^3 \, kg^{-1} \, s^{-2}$ units.

G is called the *universal gravitational constant*.

Consider, for example, the gravitational force exerted by the earth on an object of mass m on the earth's surface. Taking the mass of the earth as approximately $6 \times 10^{24} \, kg$ and its radius as $6.4 \times 10^6 \, m$ we have

$$F \approx \frac{Gm(6 \times 10^{24})}{(6.4 \times 10^6)^2} = \frac{(6.7 \times 6 \times 10^{13})m}{6.4^2 \times 10^{12}} = 9.8 \, m$$

This confirms that the weight of an object of mass m at the surface of the earth is mg.

Note that the size of the object was not taken into account. This is because, unless the object were enormous, its 'radius' would make negligible difference to its distance from the centre of the earth.

The effective radius of the earth, however, does vary a little between sea-level and the tops of high mountains. It follows from the formula for F that the weight of an object should decrease slightly as its height above sea-level increases and this is consistent with practical observation.

When assessing the motion of an object relative to the earth, or some other massive body of mass m_1, the formula $F = \dfrac{Gm_1 m_2}{r^2}$ is often simplified by combining Gm_1 into a single constant k say, so that $F = \dfrac{km}{r^2}$.

In this case the unit for k is $m^2 \, s^{-2}$.

Newton's Law of Universal Gravitation is, like all physical 'laws', no more than a model based on the evidence available at the time. With the advance of astronomical observation and the study of the motion of atomic particles, it became apparent that some errors arose when using Newton's law in situations where velocities approached the speed of light. Einstein's theory of relativity (early C20) resolved these discrepancies but, for ordinary circumstances, applying Newton's laws to the motion of moving objects is quite accurate enough.

Examples 9c

1. A satellite of mass M is orbiting the earth at a height of 2×10^5 m above the earth's surface. The force of attraction, F newtons, exerted by the earth on the satellite is given by $F = \dfrac{4 \times 10^{14} M}{r^2}$, where r metres is the distance of the satellite from the centre of the earth. By modelling the earth as a sphere of radius 6.4×10^6 m and the satellite as a particle,

(a) find the acceleration of the satellite towards the earth.

Assuming that the satellite performs a circular orbit, find

(b) its angular velocity   (c) the time taken to complete one orbit.

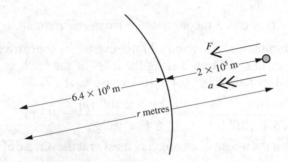

$$r = 6.4 \times 10^6 + 2 \times 10^5 = 6.6 \times 10^6$$

$$F = \frac{4 \times 10^{14} M}{r^2} = \frac{4 \times 10^{14} M}{(6.6 \times 10^6)^2} = 9.182 \ldots M$$

$$F = 9.18 M \quad (3 \text{ sf})$$

(a) Using Newton's law $F = ma$ gives

$$(9.18 \ldots) M = Ma \quad \Rightarrow \quad a = 9.182 \ldots$$

The acceleration towards the earth is $9.18\,\mathrm{m\,s^{-2}}$ (3 sf)

(b) For a particle travelling in a circle the acceleration towards the centre is $r\omega^2$ where ω is the angular velocity.

$$a = r\omega^2 \quad \Rightarrow \quad \omega^2 = \frac{9.182 \ldots}{6.6 \times 10^6}$$

The angular velocity of the satellite is $1.18 \times 10^{-3}\,\mathrm{rad\,s^{-1}}$ (3 sf)

(c) The time of one revolution is $2\pi/\omega$ seconds,

i.e. $2\pi \div (1.18 \ldots \times 10^{-3})\,\mathrm{s} \quad \Rightarrow \quad 5320\,\mathrm{s}$ (3 sf)

One orbit takes approximately 1.5 hours.

2. A rocket of mass M is fired vertically from the surface of the earth with a speed V and moves under the action of gravity only. The speed V is not great enough for the rocket to 'escape' from the earth's gravitational field.

Use the law of gravitation in the form $F = \dfrac{kM}{x^2}$ where x is the distance at any time between the rocket and the centre of the earth.

(a) Express k in terms of g and R, the radius of the earth at the launch site.

(b) Find the greatest distance from the centre of the earth reached by the rocket, giving your answer in terms of g and R.

(a) At the surface of the earth, $x = R$ and the gravitational force is Mg

$$\therefore \quad Mg = \frac{kM}{R^2} \quad \Rightarrow \quad k = gR^2$$

(b)

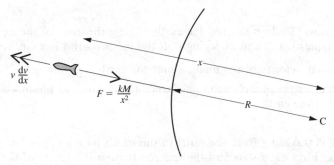

The force acting on the rocket at any time is $\dfrac{kM}{x^2} = \dfrac{gR^2 M}{x^2}$

Using $v\dfrac{dv}{dx}$ for the acceleration, we have

$$-\frac{gR^2 M}{x^2} = Mv\frac{dv}{dx}$$

$$\Rightarrow \quad -\int \frac{gR^2}{x^2}\, dx = \int v \, dv$$

$$\therefore \quad \frac{gR^2}{x} = \frac{1}{2}v^2 + A$$

When $x = R$, $v = V$ therefore $A = gR - \frac{1}{2}V^2$

$$\Rightarrow \quad \frac{1}{2}v^2 = \frac{gR^2}{x} - gR + \frac{1}{2}V^2$$

When the greatest distance D is reached, the velocity becomes zero,

$$\therefore \quad 0 = \frac{gR^2}{D} - gR + \frac{1}{2}V^2 \quad \Rightarrow \quad gR^2 = D\left(gR - \tfrac{1}{2}V^2\right)$$

$$\therefore \quad D = \frac{2gR^2}{2gR - V^2}$$

EXERCISE 9c

In this exercise take the value of G as $6.7 \times 10^{-11} \, \text{m}^3 \, \text{kg}^{-1} \, \text{s}^{-2}$ unless another instruction is given.

1. Using the method of the Cavendish experiment of 1798, the gravitational force between two lead spheres is found. The masses of the spheres are $0.008 \, \text{kg}$ and $12 \, \text{kg}$ and the distance between their centres is $0.01 \, \text{m}$. The force of attraction between them is $6.36 \times 10^{-8} \, \text{N}$. Use the data to find a value for G.

2. Use the Law of Universal Gravitation to estimate the mass of the earth by considering the gravitational force on a particle of mass $m \, \text{kg}$ at the earth's surface. Treat the earth as a sphere of radius 6.4×10^6; take the acceleration due to gravity at the surface of the earth as $9.8 \, \text{m} \, \text{s}^{-2}$.

3. The moon takes 27.3 days to orbit the earth. Using the mass of the earth calculated in question 2, and assuming that the moon's orbit is a circle, find

   (a) the angular velocity of the moon about the earth

   (b) the radius of the moon's orbit and hence the distance of the moon from the surface of the earth.

4. A spacecraft is travelling from the earth to the moon on a straight line joining their centres. Find the distance of the spacecraft from the centre of the earth when it reaches the point at which the gravitational attraction of the earth is equal to that of the moon. The mass of the earth is 81 times the mass of the moon and the distance between their centres is $3.8 \times 10^8 \, \text{m}$.

5. A spacecraft is put into a circular orbit about the moon. It takes 109 minutes to complete each orbit at a height of $5 \times 10^4 \, \text{m}$ above the surface. The radius of the moon is $1.7 \times 10^6 \, \text{m}$. Find

   (a) the radius of the orbit

   (b) the angular velocity of the spacecraft

   (c) the mass of the moon.

6. A man of mass $80 \, \text{kg}$ is standing on the surface of the moon. He accidently releases a piece of moon rock which he is studying and it drops from a height of $1.4 \, \text{m}$ and lands on his foot. Take the mass of the moon to be $7.4 \times 10^{22} \, \text{kg}$ and its radius to be $1.7 \times 10^6 \, \text{m}$. Find

   (a) the weight (in newtons) of the man on the moon

   (b) the acceleration due to the moon's gravity

   (c) the speed with which the rock hits his foot.

7. A rocket is fired vertically from the surface of the earth and attains a velocity $u\,\mathrm{m\,s^{-1}}$ near the surface. It then moves acted on by gravity only, assuming that air resistance is negligible.
After t seconds its distance from the centre of the earth is x metres and its velocity is $v\,\mathrm{m\,s^{-1}}$.
The radius of the earth is R metres.

Use the law of gravitation in the form $F = \dfrac{km}{x^2}$.

(a) Write down the equation of motion of the rocket and by integration show that

$$v^2 = u^2 + 2k\left(\frac{1}{x} - \frac{1}{R}\right)$$

(b) Find the value which v^2 approaches as x increases.

(c) Using the result of part (b) find, in terms of k and R, the value which u must exceed if the rocket is never to return to earth (i.e. find the escape velocity).

(d) Evaluate the escape velocity using $k = 4 \times 10^{14}\,\mathrm{N\,m^2\,kg^{-1}}$ and $R = 6.4 \times 10^6$.

*8. A space station is set up on Mars with the ability to launch spacecraft from the surface of that planet. A rocket is fired vertically from the surface with initial velocity $u\,\mathrm{m\,s^{-1}}$ and it reaches a greatest height h metres.
For a body of mass m kilograms the gravitational attraction of Mars is F newtons when the body is at a distance x kilometres from the centre of Mars.
The mass of Mars is M kilograms where $M = 6.4 \times 10^{23}$,
its radius is R metres where $R = 3.4 \times 10^6$
and the acceleration due to gravity at its surface is $g_1\,\mathrm{m\,s^{-2}}$

(a) find the value of g_1.

If the law of gravitation is to be used in the form $F = \dfrac{km}{x^2}$

(b) find the value of k.

(c) express k in terms of g_1 and R.

(d) find h in terms of g_1, R and u.

(e) by considering the values of u for which h is large, deduce the 'escape' velocity from Mars

    (i) in terms of g_1 and R,

    (ii) evaluated corrected to 2 significant figures.

THE SIMPLE PENDULUM

When a heavy particle, attached to one end of a light string whose other end is fixed, oscillates through a *small* angle, the system is called a *simple pendulum*.

In the following diagram we will take m as the mass of the particle P, l as the length of the string and θ as the angle between the string and the vertical at any time t. So that quantities can be marked clearly on the diagram, the angle is shown larger than it should be.

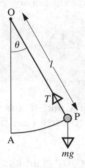 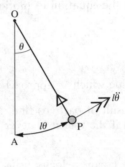

The length of the arc AP is $l\theta$, so the acceleration in the direction of the tangent is $\dfrac{d}{dt}(l\dot{\theta})$, or $l\ddot{\theta}$, away from A.

The force along the tangent is $mg \sin\theta$ towards A.

Therefore using $F = ma$ gives $mg \sin\theta = -ml\ddot{\theta}$

Now θ is at all times a small angle, so $\sin\theta \approx \theta$.

Therefore $mg\theta \approx -ml\ddot{\theta}$

which can be written $\qquad l\ddot{\theta} \approx \dfrac{-g}{l}(l\theta)$ [1]

From the diagram we see that $\quad x = l\sin\theta \approx l\theta$

$\therefore \qquad\qquad\qquad\qquad \ddot{x} \approx l\ddot{\theta}$

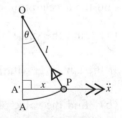

Equation [1] then becomes $\qquad \ddot{x} \approx \dfrac{-g}{l}x$

showing that the acceleration $\ddot{x}$, is proportional to the displacement and in the opposite direction.

Hence, as motion along the arc is approximately equivalent to motion along A'P,

to a good approximation the particle describes simple harmonic motion.

Comparing with the standard equation of SHM, i.e. $\ddot{x} = -n^2x$, we see that the constant n^2 is equal to $\dfrac{g}{l}$.

Therefore the period T of a complete oscillation is $\dfrac{2\pi}{n}$ where $n^2 = \dfrac{g}{l}$

i.e. $$T = 2\pi\sqrt{l/g}$$

This formula can be quoted unless its derivation is asked for.

Note that T depends on the length of the string but not on the mass of the particle (often called the *pendulum bob*).

The Seconds Pendulum

A simple pendulum which swings from one end of its path to the other end in exactly one second is called a *seconds pendulum* and is said to *beat seconds*.

Since each half oscillation takes 1 second, the period of oscillation is 2 seconds, i.e. $T = 2$.

The length of string, l, required for a seconds pendulum can then be calculated using

$$T = 2\pi\sqrt{l/g} \quad \Rightarrow \quad 2 = 2\pi\sqrt{l/g}$$

giving $l = \dfrac{g}{\pi^2}$ $(\approx 0.99\,\text{m})$

Examples 9d

1. The period of a simple pendulum is $3T$. If the period is to be reduced to $2T$, state whether the length of the pendulum should be increased or decreased and find the percentage change required.

Taking the original length of the pendulum as l, we have

$$3T = 2\pi\sqrt{\dfrac{l}{g}} \qquad\qquad\qquad\qquad [1]$$

The period increases as l increases so, to reduce the period, the length should be reduced.

Let kl be the required reduction in length.

Then $$2T = 2\pi\sqrt{\dfrac{(l-kl)}{g}}$$

$$\Rightarrow \qquad \left(\dfrac{T}{\pi}\right)^2 = \dfrac{l(1-k)}{g}$$

From [1] $\left(\dfrac{T}{\pi}\right)^2 = \dfrac{4l}{9g}$

$\therefore$ $\dfrac{4l}{9g} = \dfrac{l(1-k)}{g}$ $\Rightarrow$ $9(1-k) = 4$

$\therefore$ $k = \frac{5}{9}$ and the reduction in length is $\frac{5}{9}l$.

The percentage reduction in length is $\dfrac{kl}{l} \times 100$,   i.e. 56%   (2 sf).

2. **At a location where $g = 9.81\,\text{m s}^{-2}$ a seconds pendulum beats exact seconds. If it is taken to a place where $g = 9.80\,\text{m s}^{-2}$ by how many seconds per day will it be wrong?**

If l is the length of the pendulum then

$$\pi\sqrt{\dfrac{l}{9.81}} = 1   \text{ giving }   \sqrt{l} = \dfrac{\sqrt{9.81}}{\pi}$$

When $g = 9.80$ the time, t, of one beat is given by $\pi\sqrt{\dfrac{l}{9.8}}$

and is therefore $\sqrt{\dfrac{9.81}{9.8}}$ seconds.

The number of beats in 24 hours is now $(24 \times 60 \times 60) \div \sqrt{\dfrac{9.81}{9.8}}$

The number of beats lost in 24 hours is therefore

$$24 \times 60 \times 60\left(1 - \sqrt{\dfrac{9.8}{9.81}}\right) = 24 \times 60 \times 60\,(0.0005)$$

$$= 44$$

Therefore, where $g = 9.80$ the pendulum will lose 44 seconds per day.

EXERCISE 9d

1. A simple pendulum is 2 m in length and the time it takes to perform 50 complete oscillations is measured.

    (a) The pendulum is on the earth and the time taken is 142 s. Find g.

    (b) The pendulum is on the moon and the time taken is 341 s. Find the acceleration due to gravity on the moon.

2. Two simple pendulums have periods 1.6 s and 2.4 s. They are set oscillating, initially in step.

    (a) Find the length of each pendulum.

    (b) Find the interval after which they are next in step.

3. The length of a simple pendulum is shortened by 54 cm and this has the effect of reducing the period by 20%. Find the length of the original pendulum.

4. By what length should a simple pendulum be shortened if it is meant to beat seconds but loses 40 seconds in 12 hours?

5. A pendulum which beats seconds where $g = 9.81\,\mathrm{m\,s^{-2}}$ is taken up a mountain to a place where it loses 30 seconds per day. What is the value of g at the new location?

6. If a simple pendulum which beats exact seconds has its length changed by 1%, find the number of seconds by which it will be inaccurate if

(a) the length is increased,

(b) the pendulum is shortened.

FORCES THAT PRODUCE SIMPLE HARMONIC MOTION

When a particle performs SHM, its acceleration is always directed towards, and is proportional to, its distance from a fixed point. It follows that SHM is produced by the action of *a force directed towards a fixed point and proportional to the distance from that point.*

The tension in an elastic string is proportional to the extension and always acts to restore the string to its natural length. These are the conditions in which SHM is likely, and we are now going to consider the motion of a particle attached to the end of an elastic string.

Examples 9e

1. A particle P of mass 2 kg is lying at a point A on a smooth horizontal table. P is attached to one end of a light elastic string of length 1 m and modulus of elasticity 8 N, whose other end is fixed at a point O on the table. Initially the string is just taut. P is pulled away from O until it is 2 m from O and is then released.

(a) Show that, at first, P performs SHM and state the position of P when SHM ceases.

(b) Find the time taken to reach the point found in part (a) and the speed of P at this point.

(c) Describe briefly the motion of P after it passes through the position found in part (a) and give an assumption that is made.

(a)

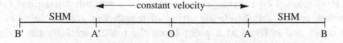

Consider the particle in a general position, distant x from A, where the string is taut.

As long as the string is taut, Hooke's Law $\left(\text{tension} = \dfrac{\lambda \times \text{extension}}{\text{natural length}} \right)$ gives $T = 8x$

Then using $F = ma$ gives $T = -2\ddot{x}$

$\therefore \qquad 8x = -2\ddot{x} \qquad \Rightarrow \qquad \ddot{x} = -4x$.

This is the equation of SHM in which $n^2 = 4$

$\therefore \qquad$ P performs SHM as long as the string is taut.

SHM ceases when the string becomes slack, i.e. when P reaches A.

(b) At A, $x = 0$ therefore A is the centre of the SHM that has been performed so far and at this stage P has covered one quarter of an oscillation from the end B to the centre A.

$\therefore \qquad$ the time taken to reach A is given by $\dfrac{1}{4} \left(\dfrac{2\pi}{n} \right)$

Therefore, as $n = 2$, the time taken is $\frac{1}{4}\pi$ seconds

The amplitude, a, is OA, i.e. 1 m.

The speed of P at the centre of the SHM is given by an, and is $2\,\mathrm{m\,s^{-1}}$.

(c) After P passes through A the string is slack so there is no horizontal force acting on it. Therefore P travels on with constant velocity $2\,\mathrm{m\,s^{-1}}$ until the string becomes taut again. This will happen when P is at a point A' on the opposite side of O such that OA = OA'. Then P will again move with SHM.

```
           ←———— constant velocity ————→
      ___SHM___                              ___SHM___
    |____|____|_____|_____|____|____|
    B'        A'           O            A         B
```

This assumes that the slack string does not interfere with the motion.

2. **A light elastic string of natural length $2a$ and modulus of elasticity λ is stretched between two points A and B, distant $4a$ apart on a smooth horizontal table. A particle of mass m is attached to the midpoint of the string and is pulled towards A through a distance a and then released. Show that the particle describes SHM; state the centre and amplitude, and find the period of the motion.**

When the particle is attached to the centre of the string, two independent strings are created, each with length a and modulus of elasticity λ. Consider the motion of the particle as it passes through a general point P, distant x ($x < a$) from M, the midpoint of AB.

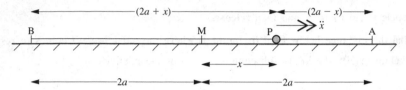

For the string BP:

natural length $= a$ and stretched length $= 2a + x$

$\therefore$ extension $= a + x$

The tension T_B is given by Hooke's Law,

i.e. $\qquad T_B = \dfrac{\lambda}{a}(a + x)$

Similarly for the string AP, for which the extension is $(2a - x) - a = a - x$

Hooke's Law gives $\quad T_A = \dfrac{\lambda}{a}(a - x)$

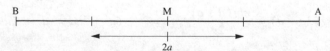

The resultant force acting on the particle is $T_B - T_A$ towards M.

Then using $F = ma$ gives

$$\frac{\lambda}{a}(a + x) - \frac{\lambda}{a}(a - x) = -m\ddot{x}$$

$$\Rightarrow \qquad\qquad \frac{2\lambda x}{a} = -m\ddot{x}$$

$$\therefore \qquad\qquad \ddot{x} = -\frac{2\lambda}{ma}x$$

This is the equation of SHM in which $n^2 = \dfrac{2\lambda}{ma}$

$$
\begin{array}{ccc}
\text{B} & \text{M} & \text{A} \\
\vdash & \vdash & \dashv
\end{array}
$$

$$\longleftarrow 2a \longrightarrow$$

$\ddot{x} = 0$ when $x = 0$ and this is at M, so M is the centre of oscillation.

The length of the path is $2a$, so the amplitude is a.

The period of the motion is $\dfrac{2\pi}{n}$, i.e. $2\pi\sqrt{\dfrac{ma}{2\lambda}}$

3. A particle of mass m is attached to one end of an elastic string of natural length $2a$ and modulus of elasticity λ. The other end of the string is fixed to a point A and when the particle hangs in equilibrium at E the string is of length $3a$.

(a) Find λ in terms of m and g.

The particle is raised to A and then released.

(b) Find the distance below E of the point B where the particle next comes to rest.

(c) Explain briefly the likely subsequent motion of the particle.

(a) The particle is in equilibrium

$$\therefore \quad T = mg$$

Using Hooke's Law gives

$$T = \frac{\lambda a}{2a} = \frac{1}{2}\lambda$$

$$\therefore \quad \lambda = 2mg$$

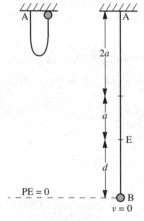

(b) Let d be the distance of B below E. To use conservation of mechanical energy we consider PE, KE and EPE

$$\left(\text{EPE} = \frac{\lambda x^2}{2 \times \text{natural length}} \right)$$

Total mechanical energy at A is

$$mg(3a+d)+0+0 = mg(3a+d)$$

Total mechanical energy at B is

$$0+0+\frac{\lambda(d+a)^2}{4a} = \frac{mg(d+a)^2}{2a}$$

Conservation of mechanical energy from A to B gives

$$mg(3a+d) = \frac{mg(d+a)^2}{2a} \quad \Rightarrow \quad 6a^2 + 2ad = d^2 + 2ad + a^2$$

$$\therefore \quad d^2 = 5a^2 \quad \Rightarrow \quad d = a\sqrt{5}$$

B is distant $a\sqrt{5}$ below E.

(c) When P is at E the tension is equal to the weight of P.
When P is above E the tension is less than the weight of P.
When P is below E the tension is greater than the weight of P.

Therefore the resultant force is always directed towards E
and depends on the extension. Therefore it seems likely
that P begins to perform SHM with centre E and
amplitude $a\sqrt{5}$.
However, when P reaches point C, where EC $= a$, the
string ceases to be stretched. The only force then acting
on the particle is its own weight so the motion of P will
be in two parts: SHM below C and motion under gravity
above C.

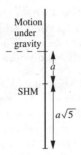

4. **A firm making small fine porcelain ornaments, wishes to ensure that the packaging used for exporting the goods is completely satisfactory. To test for the effect of vibration, a typical package is placed on a belt that is kept taut by passing round two pulleys as shown.**

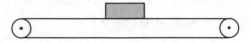

The belt can be made to oscillate along its length with simple harmonic motion performing 4 oscillations per second. If the mass of the package is 0.6 kg and the coefficient of friction with the belt is 0.9, find the greatest amplitude permissible if the package must not slip on the belt.

Four oscillations are performed per second, therefore the period of the motion is
0.25 seconds

The period is given by $\dfrac{2\pi}{n}$ therefore $\dfrac{2\pi}{n} = 0.25$ $\Rightarrow$ $n = 8\pi$

The maximum acceleration of the belt occurs at each end of the oscillation and
its magnitude is n^2a, where a metres is the amplitude.

$\therefore$ the magnitude of the greatest acceleration is $(8\pi)^2a$

If the package is not to slip, it must move with the same acceleration as the belt,
so the maximum acceleration of the package is also $(8\pi)^2a$.

The force that must produce this acceleration is the frictional force exerted by the
belt on the package.

Using $F = m\ddot{x}$ gives $F = 0.6 \times (8\pi)^2a$

The greatest value of F is μR, i.e. μmg.

When F has this value $\mu mg = 0.6 \times (8\pi)^2a$

i.e. $0.9 \times 0.6 \times 9.8 = 0.6 \times 64\pi^2a$

$\Rightarrow$ $a = 0.013\,96\ldots$

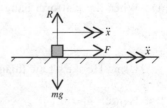

Therefore the amplitude must not exceed 1.40 cm (3 sf)

5. The diagram shows a fairground 'Test your
   Strength' machine consisting of a long spring fixed
   at one end A, and with a platform of mass 2 kg
   attached to the other end.

   Competitors strike the platform vertically
   downwards with a mallet, as hard as they can. The
   depth to which the platform descends is recorded on
   a scale and the person who attains the greatest
   depth wins. The natural length of the spring is 2 m
   and its modulus of elasticity is $40g$ N. The winner
   causes the platform to descend by 0.5 m.

   By modelling the platform as a particle,

   (a) find the length of the spring when the
        platform hangs freely at E,

   (b) find the initial speed of the platform when
        struck (use $g = 10$),

   (c) find the impulse of the blow from the mallet,

   (d) show that the particle describes simple harmonic motion.

   State any assumptions that have been made and suggest any way in which you think
   the model might be improved.

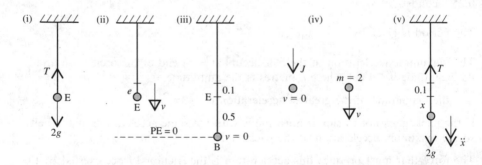

   (a) When the platform hangs at rest, (diagram i)

   $$T = 2g$$

        Using Hooke's Law (diagram ii) gives $$T = 40g \frac{e}{2}$$

        Hence $e = 0.1$

        The length of the spring is 2.1 m

(b) Conservation of mechanical energy can be used from E (ii) to B (iii).

At E, $PE = 2g(0.5)$ $KE = \dfrac{1}{2}(2)v^2$ $EPE = \dfrac{40g}{4}(0.1)^2$

At B, $PE = 0$ $KE = 0$ $EPE = \dfrac{40g}{4}(0.5)^2$

Conservation of ME gives

$$g + v^2 + 0.1g = 3.6g \qquad \Rightarrow \qquad v^2 = 2.5g$$

$$\Rightarrow \qquad v = 5 \quad (g = 10)$$

The speed at E is $5\,\mathrm{m\,s^{-1}}$

(c) Taking $J\,\mathrm{N\,s}$ as the impulse and using   impulse $=$ increase in momentum
(diagram iv) we have

$$J = 2v \qquad \Rightarrow \qquad J = 10$$

The impulse is $10\,\mathrm{N\,s}$

(d) When the platform is in a general position P, where the depth below E is x
(diagram v), using Newton's Law gives

$$2g - T = 2\ddot{x}$$

Hooke's Law gives $T = \dfrac{40g}{2}(0.1 + x)$

$\therefore$ $2g - (2g + x) = 2\ddot{x}$

$\Rightarrow$ $\ddot{x} = -x$

This is the equation of SHM (for which $n = 1$ and about E as centre)

The *spring* can push as well as pull so the equation $\ddot{x} = -x$ applies in all
positions so in theory the platform performs SHM throughout the whole of
its path.

Assumptions are:

The spring is light and cannot distort.

The machine has no device for 'damping down' the motion as the platform
descends.

There is no air resistance to the motion of the platform.

Possible improvements:

The surface area of the platform does offer some air resistance so a term
reflecting this could be incorporated into the equation of motion.

EXERCISE 9e

In questions 1 to 4 a particle P is moving with SHM on a smooth horizontal surface under the action of a horizontal force. The path lies between A and B, and O is the midpoint of AB.

1. If the mass of P is 2 kg, the amplitude of the motion is 5 m and the period is $\frac{2\pi}{3}$ s, find the magnitude of the horizontal force acting on P

   (a) when P is 4 m from O

   (b) when the horizontal force has its greatest magnitude.

2. When P is 0.32 m from O its speed is 3.6 m s^{-1} and the horizontal force acting on it is 18 N. Given that the mass of P is 0.25 kg find

   (a) the period   (b) the amplitude of the motion.

3. The mass of P is 0.2 kg and it is projected from O towards A with speed 3 m s^{-1}. The motion takes place under the action of a horizontal force of $20x$ newtons, directed towards O, where x metres is the distance of P from O. Find

   (a) the period

   (b) the amplitude

   (c) the magnitude of the maximum horizontal force on P.

4. When P is at A, the horizontal force acting on it is 90 N towards O. After P has moved a distance of 2 m towards B, this force is 54 N.
   If the mass of P is 0.5 kg find

   (a) the period   (b) the amplitude.

5. The motion of the piston in a car engine is approximately SHM with an amplitude of 5 cm. The mass of the piston is 0.5 kg.
   When the piston is making 60 oscillations per second find

   (a) the maximum force required to give the piston this motion

   (b) the maximum speed of the piston.

6. A particle P of mass 2 kg is placed on a rough horizontal surface. The coefficient of friction between the particle and the surface is 0.4. The surface moves horizontally with SHM of period T seconds and amplitude a metres. Find whether the particle will slip on the surface

   (a) if $T = 4$ and $a = 1.2$ (b) if $T = 2$ and $a = 0.7$.

7. A fairground ride consists of a rough horizontal surface which is made to oscillate horizontally with approximate SHM of amplitude 0.5 m. A girl of mass 50 kg stands on the ride. The coefficient of friction between the girl and the surface is 0.3. Find the greatest possible value of the period of oscillation if the girl is not to slip.

8. A particle P of mass m is placed on a horizontal surface which oscillates vertically up and down in SHM with amplitude a and period $\dfrac{2\pi}{\omega}$. Take an origin O at the centre of the motion and an x-axis with positive direction vertically upwards. When the displacement of P from O is x, the normal reaction is R.

   (a)   Write down the equation of motion.

   (b)   Find R in terms of m, ω and x.

   (c)   Find (i)   the least value of R (ii)   the greatest value of R.

   (d)   Find the greatest amplitude the motion can have if P is to remain in contact with the platform.

9. During a storm a boat is riding up and down on the waves in approximate SHM of amplitude 4.9 m. At one extreme position of the motion the normal reaction exerted on a passenger of mass 60 kg, by his seat, has half the value it would have if the boat were stationary. Find

   (a)   the period of an oscillation

   (b)   the normal reaction from the seat when the boat is at the opposite extreme position.

10. A particle P, of mass 0.5 kg, is lying on a smooth horizontal table and is attached to one end of a light elastic spring of natural length 0.8 m and modulus of elasticity 40 N. The other end of the spring is fixed to a point A on the table. P is held 0.6 m from A, thus compressing the spring, and is then released.

   (a)   Show that P performs SHM and find the period.

   (b)   Find the distance AP when P first comes to instantaneous rest.

   (c)   Will P perform complete oscillations in SHM?

11. A particle P, of mass 2 kg is lying on a smooth horizontal surface. P is attached to a point A on that surface by a light elastic string of modulus 18 N and natural length 1 m. Initially P lies at rest at point B, which is 1 m from A. P is then projected in the direction AB with velocity $1.8 \, \mathrm{m\,s^{-1}}$.

   (a)   Show that, at first, P performs SHM.

   (b)   Find the distance AP when P first comes to instantaneous rest.

   (c)   Find the time taken for P to return to B.

   (d)   Find the velocity of P when (i)   AP $= 1.4$ m   (ii)   AP $= 0.6$ m.

12. A particle P of mass 0.4 kg is lying on a smooth horizontal surface. P is attached
to a point A on that surface by a light elastic spring of modulus 5 N and natural
length 0.5 m. Initially P lies at rest at point B, which is 0.5 m from A. P is then
projected in the direction AB with velocity u metres per second.

(a) Use conservation of energy to find the length AP when P is in any possible
position of instantaneous rest   (i) if $u = 1$ (ii) if $u = 3$.
State in each case whether complete SHM oscillations are possible and also
any assumptions which you need to make.

(b) Where complete oscillations are possible, find the period and amplitude.

13. Two points A and B are distant $5l$ apart on a smooth horizontal table.
A particle P, of mass m, is attached to A by an elastic string of natural length l
and modulus of elasticity mg and to B by an elastic string of natural length l and
modulus of elasticity $2mg$.
Initially P lies in equilibrium at point O on the line AB.

(a) Find the lengths of AO and OB.

P is then moved along the line AB until   AP $= 2l$ and is released from rest at
that point.

(b) Taking O as the origin and an x-axis in the direction OA, consider the
particle when its displacement from O is x metres. (Put P on the positive
part of the x-axis.)
  (i)   Find in terms of x the extensions in AP and BP.
  (ii) Write down the equation of motion and, by simplifying it, deduce
        that P performs SHM with centre at O.
  (iii) Find the period and amplitude.

14. A particle of mass 0.5 kg is attached to one end of a light elastic string of natural
length 2 m and modulus of elasticity 7 N. The other end of the string is fixed to
a point A and the particle hangs in equilibrium at a point E.

(a) Find the distance AE.

The particle is then pulled down a further distance a metres and released from
rest.

(b) Taking E as the origin and an x-axis vertically downwards, consider the
particle when it is at a distance x below E.
  (i)   Write down the equation of motion and, by simplifying it, deduce
        that P performs complete oscillations of SHM with centre at E,
        provided that $a \leqslant 1.4$.
  (ii) Find the period.
  (iii) Describe the motion if $a = 1.8$.

15. A particle of mass $2m$ is attached to one end of a light elastic string of natural length l. The other end of the string is fixed to a point A and the particle hangs in equilibrium at a point E, where $AE = 2l$. It is then projected vertically downwards from E with an initial velocity $\sqrt{gl}$.

    (a)   Find the modulus of elasticity.

    (b)   Use conservation of energy to find the depth below A at which the particle next comes to instantaneous rest.

    (c)   Show that the particle performs SHM and find the period and amplitude.

    (d)   Describe the motion if the initial velocity is changed to a value
        (i)   less than $\sqrt{gl}$ (ii)   greater than $\sqrt{gl}$

*16. 'Springmakers' Ltd have been asked to provide a spring, of length $30\,\text{cm}$, for an application where it is required to make a mass of $0.5\,\text{kg}$ oscillate vertically at a rate of 2 oscillations per second.

    (a)   Find the modulus of elasticity which the spring should have.

When the spring is fitted the rate of oscillation is found to be 2.2 oscillations per second. It is decided to correct the rate to the required 2 oscillations per second by attaching an extra, separate mass to the end of the spring.

    (b)   Find the extra mass needed to make this adjustment.

CONSOLIDATION C

SUMMARY

Variable Motion

Take s, v, a and t to represent displacement, velocity, acceleration and time, when a, v or s is a function of time, $f(t)$, use:

$$\dot{s} = v = \frac{ds}{dt} \qquad \dot{v} = a = \frac{dv}{dt}$$

$$v = \int a \; dt \qquad s = \int v \; dt$$

When a is a function of displacement, $f(s)$, use:

$$a = v\frac{dv}{ds} \quad \Rightarrow \quad \int f(s) \; ds = \int v \; dv \quad \text{(giving } v \text{ as a function of } s\text{)}$$

When v is a function of displacement, $f(s)$, use:

$$a = v\frac{dv}{ds} = f(s) \times \frac{df(s)}{ds}$$

$$v = \frac{ds}{dt} = f(s) \quad \Rightarrow \quad \int \frac{1}{f(s)} \; ds = \int 1 \; dt$$

Simple Harmonic Motion

SHM is motion in a straight line in which the acceleration is proportional to the distance from a fixed point on the line, O say, and is always directed towards that point.

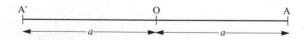

AA′ is the path

O is the centre, or mean position

a is the amplitude (the distance OA)

An oscillation is the journey from A to A′ and back to A.

T is the period of an oscillation

For a general position of the particle P where $\quad OP = x$,

$$\ddot{x} = -n^2 x \quad \text{where } n \text{ is a constant}$$

$$\dot{x} = -n\sqrt{(a^2 - x^2)}$$

$$x = a \cos nt \quad \text{where} \quad t = 0 \quad \text{when} \quad x = a$$

$$T = \frac{2\pi}{n}$$

The maximum acceleration occurs at A and A′ and its magnitude is $n^2 a$.

The maximum speed is na, occurring at O.

Associated Circular Motion

As a point P travels round a circle at constant angular speed ω, its projection on a diameter of the circle describes SHM with equation

$$\ddot{x} = -\omega^2 x$$

The Simple Pendulum

When a heavy particle, attached to one end of a light string whose other end is fixed, oscillates through a *small* angle, the system is called a simple pendulum. To a good approximation the particle describes simple harmonic motion.

The period T of a complete oscillation, i.e. a swing forward and back, is given by $\quad T = 2\pi\sqrt{l/g}$.

A 'seconds pendulum' is designed to take exactly 1 second to swing through half an oscillation.

Forces Producing Simple Harmonic Motion

SHM is produced by the action of a force that is directed towards a fixed point and is proportional to the distance from that point. The commonest example of such a force is the tension in a stretched elastic string acting on a particle attached to the end of the string.

Variable Forces

When the force acting on an object of mass m is a function of time, i.e. $F = f(t)$, Newton's Law of Motion can be applied using $a = dv/dt$ giving

$$f(t) = m\frac{dv}{dt} \quad \Rightarrow \quad \int f(t)\ dt = \int m\ dv$$

When the force is a function of displacement, i.e. $F = f(s)$, Newton's Law of Motion can be applied using $a = v\ dv/ds$ giving

$$f(s) = mv\frac{dv}{ds} \quad \Rightarrow \quad \int f(s)\ ds = \int mv\ dv$$

When the point of application of a force F, where $F = f(s)$, moves through a displacement s, the *work done* is given by $\int f(s)\ ds$.

If the PE does not change then $\int f(s)\ ds = \frac{1}{2}mv^2 - \frac{1}{2}mu^2$.

The impulse exerted over an interval of time t, by a force F where $F = f(t)$, is given by $\int f(t)\ dt$. It follows that $\int f(t)\ dt = mv - mu$.

Law of Universal Gravitation

This law states that if the centres of any two bodies in the universe are at a distance r apart and their masses are m_1 and m_2, the force F that attracts each to the other is given by

$$F = \frac{Gm_1 m_2}{r^2}$$

The value of the constant G is approximately 6.7×10^{-11} and G is measured in $m^3\,kg^{-1}\,s^{-2}$ units.

MISCELLANEOUS EXERCISE C

In questions 1 to 4 a problem is set and is followed by a number of suggested responses. Choose the correct response.

For questions 1 and 2, use this diagram of a particle P moving along a straight line.

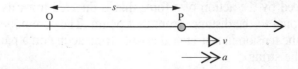

1. Given that $v = s^2$, the acceleration is given by

   A $2s$ **B** 2 **C** $2s^3$ **D** $\frac{1}{3}s^3$

2. If $v = e^{2t}$ and $s = 0$ when $t = 0$, the value of s at any time t is given by

   A $\dfrac{1}{t}e^{2t}$ **B** $\frac{1}{2}(e^{2t} - 1)$ **C** $2(e^{2t} - 1)$ **D** $\frac{1}{2}e^{2t}$

3. A particle moves along a straight line with an acceleration of $\dfrac{2}{v}$ where v is the velocity at any instant. Initially the particle is at rest. The velocity of the particle at time t is:

   A $2t$ **B** $4t$ **C** $\dfrac{2t}{v}$ **D** $2\sqrt{t}$

4. A particle moves in a straight line with an acceleration $2s$ where s is its displacement from a fixed point on the line, and $v = 0$ when $s = 0$. Its speed when its displacement is s is:

   A s **B** s^2 **C** $s\sqrt{2}$ **D** $\sqrt{2s}$

5. A particle P moves in a straight line Ox so that at time t seconds, its acceleration in the direction of increasing x is $8te^{2t}\,\mathrm{m\,s^{-2}}$. Given that P starts from rest at O when $t = 0$, find

   (a) the speed of P when $t = \frac{1}{2}$,

   (b) the distance of P from O when $t = \frac{1}{2}$.   (ULEAC)$_\mathrm{s}$

6. A particle P, of mass $2\,\mathrm{kg}$, is moving under the influence of a variable force **F**. At time t seconds, the velocity $\mathbf{v}\,\mathrm{m\,s^{-1}}$ of P is given by

$$\mathbf{v} = 2t\mathbf{i} + e^{-t}\mathbf{j}.$$

   (a) Find the acceleration, $\mathbf{a}\,\mathrm{m\,s^{-2}}$, of P at time t seconds.

   (b) Calculate, in N to 2 decimal places, the magnitude of **F** when $t = 0.2$.
                                                    (ULEAC)

7. A dot moves on the screen of an oscilloscope so that its position relative to a fixed origin is given by:

$$\mathbf{r} = 2t\mathbf{i} + \sin\left(\frac{\pi t}{2}\right)\mathbf{j}$$

   (a) Sketch the path of the dot for $0 \leqslant t \leqslant 4$.

   (b) Find the velocity and acceleration of the dot when $t = 3$. Draw vectors on your diagram to show these two quantities.   (AEB)

8. The position vector **r** metres of a particle P at time t seconds is given by

$$\mathbf{r} = (\cos 2t)\mathbf{i} - (\sin 2t)\mathbf{j}.$$

(a) Find the velocity of P at time t seconds.

(b) Show that the speed of P is constant and find its value.   (ULEAC)

9. A particle P, of mass 0.2 kg, moves in a straight line through a fixed point O. At time t seconds after passing through O, the distance of P from O is x metres, the velocity of P is $v\,\mathrm{m\,s^{-1}}$ and the acceleration of P is $(x^2 + 4)\,\mathrm{m\,s^{-2}}$.

(a) Use the information given to form a differential equation in the variables v and x only for the motion of P.

Given that $v = 2$ when $x = 0$,

(b) show that $3v^2 = 2x^3 + 24x + 12$.

(c) Find, in J, the work done on P by the force producing its acceleration as P moves from $x = 0$ to $x = 9$.   (ULEAC)

10. A steel ball of mass 0.1 kg falls vertically through thick oil and, in addition to a constant gravitational force, it is subject to a resistance, the magnitude of which is $49v\,\mathrm{N}$, where $v\,\mathrm{m\,s^{-1}}$ is the speed of the ball.

(a) At time $t = 0$, the ball is released from rest.
   (i) Show that the speed of the ball at time t is given by

$$v = 0.02(1 - e^{-490t}).$$

   (ii) Draw a speed–time graph and discuss the motion for large values of t.

(b) Describe the motion when the ball is projected downwards with speed $0.02\,\mathrm{m\,s^{-1}}$.   (AEB)$_s$

11. An aeroplane of mass M kg moves along a horizontal runway, starting from rest. The aeroplane's engines exert a constant thrust of T newtons and, when the speed of the aeroplane is $v\,\mathrm{m\,s^{-1}}$, the magnitude of the resistance to motion is kv^2 newtons, where k is a positive constant.

Show that, to reach a speed of $V\,\mathrm{m\,s^{-1}}$ on the runway, the aeroplane travels a distance

$$\frac{M}{2k} \ln\left(\frac{T}{T - kV^2}\right) \text{ metres.}$$ (ULEAC)

12. A child drops an air-filled balloon of mass 30 grams from a bridge 6 metres above a river. The balloon is slowed by air resistance of magnitude R where $R = 0.6v$ newtons and $v\,\mathrm{m\,s}^{-1}$ is the speed of the balloon at time, t. Using $g = 10$ and assuming that the upthrust of the air can be neglected,

(a) write down an equation of motion for the balloon's descent;

(b) hence show that the terminal speed of the balloon is $0.5\,\mathrm{m\,s}^{-1}$;

(c) show that an expression for the velocity in terms of time is given by

$$v = \frac{1 - e^{-20t}}{2};$$

(d) find how long it takes for the balloon to reach half its terminal speed. $(\mathrm{SMP})_s$

13. The engine of a train of total mass M kilograms generates P watts of power. Resistance to motion is Mkv^2 newtons where v is the speed of the train in metres per second and k is a positive constant. The train starts from rest and continues to accelerate under full power.

Find

(i) its terminal speed,

(ii) an expression for the train's acceleration in terms of M, P, k and v,

(iii) an expression for its speed v in terms of the distance x travelled from its initial position. (MEI)

14. A particle of mass m moves along the positive x-axis under a single force, directed towards the origin O and of magnitude $\dfrac{km}{x}$, where k is a constant. The points A and B lie on the positive x-axis at distances a and $3a$ from O, respectively. Show that the work done by the force as the particle moves from A to B is

$$-km \ln 3.$$

Given that the particle has speed $2u$ at A and speed u at B, express u^2 in terms of k. (NEAB)

15. A space ship S, of mass M, near the moon, experiences a gravitational force, of magnitude F, which is directed towards O, the centre of the moon. It is known that

$$F = \frac{Mk}{r^2},$$

where $\mathrm{OS} = r$ and k is a constant. By modelling the moon as a sphere of radius $2 \times 10^6\,\mathrm{m}$ and by taking the acceleration due to gravity at the moon's surface to be of magnitude $1.6\,\mathrm{m\,s}^{-2}$,

(a) find the value, in $\mathrm{m}^3\,\mathrm{s}^{-2}$, of k.

The space-ship S moves around the moon in a circular orbit, centre O and radius $3 \times 10^6\,\mathrm{m}$.

(b) Estimate the speed of S, giving your answer in $\mathrm{m\,s}^{-1}$. $(\mathrm{ULEAC})_s$

16. The magnitude of the gravitational force between two uniform spherical bodies of mass M and m with centres A and B respectively is

$$\frac{GMm}{r^2}$$

where $r = AB$ and $G = 6.67 \times 10^{-11}\,\mathrm{m^3\,kg^{-1}\,s^{-2}}$ is the universal gravitational constant.

(a) Given that the moon is a uniform sphere of mass $7.36 \times 10^{22}\,\mathrm{kg}$ and radius $1.74 \times 10^6\,\mathrm{m}$ find, to 2 decimal places, the magnitude of the acceleration due to gravity on the surface of the moon.

(b) Deduce that an astronaut, weighing 750 N on the surface of Earth, will weigh approximately 124 N on the surface of the moon.   (AEB)ₛ

17. Assume that the gravitational attraction of the earth on an object of mass m at a distance r from the centre of the earth is $\dfrac{km}{r^2}$, where k is a positive constant.

A rocket is launched from the earth's surface, and it travels vertically upwards. When the fuel is exhausted, the distance of the rocket from the centre of the earth is a and the speed of the rocket is u. Some time later, the distance of the rocket from the centre of the earth is x and the speed of the rocket is v. Neglecting any forces other than the gravitational attraction of the earth, find an expression for v.

Deduce that, if $u^2 \geqslant \dfrac{2k}{a}$, the rocket will never fall back to the earth.

                                                                (UCLES)ₛ

In questions 18 to 23 a problem is set and is followed by a number of suggested responses. Choose the correct response.

18. A particle P describes SHM of amplitude 1 m. In performing one complete oscillation, P travels a distance:

   A 2 m   **B** 0   **C** 4 m   **D** −2 m

19.

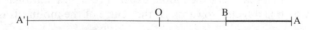

A particle travels between A and A′ with SHM of period 24 seconds. O is the centre and B is the midpoint of AO. The time taken to travel from A to B is

   A 3 s   **B** 8 s   **C** 6 s   **D** 4 s

In questions 20 and 21 a particle is moving in a straight line AA′ with SHM. The equation of motion is $\ddot{x} = -4x$ and the amplitude of the motion is 3 m.

20. The greatest speed, in $m\,s^{-1}$, is

   A 36   **B** 6   **C** 12   **D** 18

21. The time, in seconds, taken to travel from O to A is

   A $\frac{1}{8}\pi$ **B** π **C** $\frac{1}{2}\pi$ **D** $\frac{1}{4}\pi$

22. The period of a simple pendulum at a place where $g = 9.8\,m\,s^{-2}$ is 3 s. This pendulum is taken to another planet and, there, its period is 6 s. The value of g, in $m\,s^{-2}$, on that planet is

   A $\frac{1}{2}(9.8)$ **B** $(9.8)^2$ **C** $\frac{1}{4}(9.8)$ **D** $2(9.8)$

23. If a man has weight W on the surface of a spherical planet of radius R, then his weight at a height R above the surface of that planet is

   A $4W$ **B** $\frac{1}{2}W$ **C** $\frac{1}{4}W$ **D** $2W$

In each question from 24 to 29 a statement is made. Decide whether the statement is true (T) or false (F).

24. A particle whose acceleration is proportional to its displacement from a fixed point is moving with SHM.

25. A particle hanging at the end of an elastic string is pulled down and then released. The motion of the particle must be entirely SHM.

26. A particle describing linear SHM on a path AB with midpoint O has its greatest acceleration at either A or B.

27. The work done by any force F in moving an object a distance d is Fd.

28. A particle which is oscillating is not necessarily performing SHM.

29. A particle is moving along a straight line with variable acceleration. If, at some instant, the particle has a maximum velocity, the acceleration at that instant is zero.

30. A particle P, of mass 0.3 kg, moves in a horizontal straight line with simple harmonic motion of period 2 s and maximum speed $4\,m\,s^{-1}$. The centre of the path is O and the point A, on the path of P, is $\dfrac{2}{\pi}$ m from O.

Find

(a) the speed of P as it passes through A,

(b) the magnitude of the force acting on P as it passes through A.   (ULEAC)$_s$

31. A particle P describes SHM centre O, period $2\pi/3$ seconds and its maximum speed is $24\,\text{m s}^{-1}$. Find

   (i)   the amplitude of the motion,

   (ii) the time taken for P to travel from O directly to a point $4\,\text{m}$ from O.

   Given that the particle is of mass $0.25\,\text{kg}$, find

   (iii) the rate at which the force acting on P is working when $t = \pi/9$.

                                                                 (WJEC)$_s$

32. The motion of the top of a piston is modelled as Simple Harmonic with a period of $0.1\,\text{s}$ and an amplitude $0.2\,\text{m}$ about a mean position A.

   (i)   Show that x, the displacement of the top of the piston from A after t seconds, is given by the equation $x = 0.2 \sin(20\pi t)$, given that $x = 0$ when $t = 0$.

   Where appropriate your answers to the following questions should be expressed in terms of π.

   (ii) What is the greatest piston speed?

   (iii) What is the greatest magnitude of the acceleration of the piston?

   (iv) For what fraction of the period is $x > 0.1$?   (MEI)

33. A simple pendulum has a period of 1 second. It is tested in two towns, A and B. It has 3601 oscillations in an hour in town A and 3599 oscillations in an hour at town B.

   (a)   Compare the values of g for each of the two towns.

   (b)   Give a possible explanation for the difference in g between the two towns.
                                                                 (NEAB)

34. A particle P, of mass $0.01\,\text{kg}$, moves along a straight line with simple harmonic motion. The centre of the motion is the point O. At the points L and M, which are on opposite sides of O, the particle P has speeds of $0.09\,\text{m s}^{-1}$ and $0.06\,\text{m s}^{-1}$ respectively and $2OL = OM = 0.02\,\text{m}$.

   (a)   Show that the period of this motion is $2\pi \sqrt{\left(\frac{1}{15}\right)}\,\text{s}$.

   Find

   (b)   the greatest value of the magnitude of the force acting on P, giving your answer to 2 significant figures,

   (c)   the time for P to move directly from L through O to M, giving your answer to 2 significant figures.   (ULEAC)

35. A light elastic string, of natural length L and modulus $2mg$, has one end attached to a fixed point A. A particle of P, of mass m, is attached to the other end of the string and hangs freely in equilibrium at the point O, vertically below A.

(a) Find, in terms of L, the length OA.

The particle P is pulled down a vertical distance h below O, and released from rest at time $t = 0$. At time t, the displacement of P from O is x.

(b) Show that, while the string is taut,

$$\frac{d^2x}{dt^2} = -\frac{2gx}{L}.$$

(c) State the set of values of h for which P performs complete simple harmonic oscillations.

Given that $h = \frac{1}{3}L$,

(d) find the time at which P first comes instantaneously to rest,

(e) find the greatest speed of P. (ULEAC)

36. A spaceship of mass 10^4 kg is in a circular orbit 10^6 m above the surface of a planet whose diameter is 4×10^6 m.
The mutual force of attraction between the spaceship and the planet can be written as $\dfrac{k}{r^2}$, where k is a constant with units $N\,m^2$ and r is the distance in metres between the centre of the planet and the spaceship.

(a) The acceleration due to gravity on the planet's surface is $1.5\,m\,s^{-2}$. Show that $k = 6 \times 10^{16}$.

(b) Find

(i) the period of the orbit

(ii) the speed $v\,m\,s^{-1}$ of the spaceship.

(c) The spaceship uses its engines to escape from its orbit about the planet. The engines apply an impulse of magnitude $I\,N\,s$ at angle α to the spaceship's present direction of motion, as illustrated.

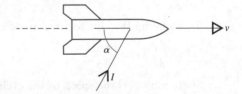

The result is that the speed of the spaceship is doubled and its new direction of motion is at an angle θ to its original direction of motion.

(i) Show that $I = 10^4 v\sqrt{(5 - 4\cos\theta)}$.

(ii) Hence find the minimum value of I as θ varies.

(iii) State the direction of the resulting velocity, in the case when $I = 10^4 v\sqrt{3}$ (UODLE)

37. A particle of mass m falls from rest, under gravity, in a medium in which the resistance to its motion is mkv, where k is a constant and v is the speed of the particle. Write down the equation of motion for the particle. If the motion were to continue indefinitely, v would approach a constant value V.

Show that $k = \dfrac{g}{V}$. Hence show that $\dfrac{dv}{dt} = \dfrac{g}{V}(V - v)$.

Show that the particle is moving with velocity $\dfrac{V}{2}$ after a time $\left(\dfrac{V}{g}\right)\ln 2$.

Show also that, during this time interval, the particle has fallen a distance s given by

$$\frac{V}{g}\int_0^{\frac{V}{2}} \frac{v}{V - v}\, dv$$

Hence show that

$$s = \frac{V^2}{g}\left(\ln 2 - \frac{1}{2}\right)$$

Find the average speed, in the form λV, during this time interval, expressing λ correct to 2 decimal places.   (NEAB)

38. A cyclist is travelling along a road.

(a) What can be deduced about the resultant force on the cycle if they are
· travelling at top speed

(i) along a straight road,

(ii) along a winding road?

A cyclist whose maximum rate of working is 600 W can reach a top speed of $10\,\text{m}\,\text{s}^{-1}$ on a level road. The combined mass of the cycle and cyclist is 90 kg.

(b) By assuming that the resistance forces of the cycle and cyclist are proportional to their speed, find a simple model for the total resistance force.

(c) By assuming that the forward force on the cyclist is constant, show that

$$\frac{dv}{dt} = -\frac{10 - v}{15}$$

where $v\,\text{m}\,\text{s}^{-1}$ is the speed of the cyclist at time t seconds.

(d) Find an expression for the speed of the cyclist in terms of time, if the cyclist starts at rest.

(e) Criticise your model for the resistance on the cyclist.   (AEB)

ANSWERS

CHAPTER 1

Exercise 1a – p. 5
1. 10 N   2. 5.14 N
3. 0.6 m
4. (a) 2.88 m   (b) 2.24 m   (c) 1.92 m
5. (a) 0.45 m   (b) 0.27 m   (c) 0.72 m
6. (a) 7.2   (b) 3.6   (c) 1.8
7. (a) 1.38 m   (b) 1.1 m   (c) 0.688 m
8. (a) 10 N   (b) $15 \, \text{m s}^{-2}$
    (c) (i) 4 N; $10 \, \text{m s}^{-2}$
       (ii) 2 N; $5 \, \text{m s}^{-2}$ (iii) 0; 0
9. 150 N
10. Ben: 780 N; Tony: 810 N; Tony is the stronger.

Exercise 1b – p. 10
1. 4.70   2. 0.823 m
3. 4.90   4. 58.8
5. 0.316 m   6. 1.43
7. 1.5 kg   8. $\frac{16}{7} a$
9. 0.5 m   10. $\frac{5}{3} a$
11. 21.1 N   12. (a) 30° (b) $\frac{1}{2} mg$
13. (a) $\frac{1}{2} mg$ (b) $\frac{1}{2} mg \sqrt{3}$ at 30° to vertical
    (c) $\frac{3}{2} mg$
14. (a) $1 - x$ (b) $\frac{2}{3}$ (c) 2.95 kg

Exercise 1c – p. 13
1. (a) 6 N   (b) $50 \, \text{m s}^{-2}$
2. 196 N
3. (a) 3.3 cm (b) 6.5 N
4. (a) 1440 N (b) 0.24 m
5. (a) 39.5 cm (b) 21.3 cm (c) 11.2 cm
6. (a) (i) 35.6 N m
     (ii) 41.2 N m; leg is uniform.
    (b) (i) 76.1 N m
     (ii) 98.1 N m; ankle cuff is 70 cm from joint.
7. (a) 588 N   (b) 1176 N
8. (a) $\frac{40}{9}$ N   (b) $\frac{160}{9}$ N; force exerted by hand acts at 9 cm from hinge, and is parallel to the spring.
9. (a) 84.5 N   (b) 666 N

CHAPTER 2

Exercise 2a – p. 18
1. (a) 1.91   (b) 10.5
2. 0.00175
3. (a) 6.94×10^{-4} (b) 7.27×10^{-5}

4. (a) 1.2   (b) 12.5   (c) 4
5. $8.38 \, \text{m s}^{-1}$ 6. $1680 \, \text{km h}^{-1}$
7. $0.785 \, \text{m s}^{-1}$
8. (a) $1.6 \, \text{rad s}^{-1}$ (b) $1.6 \, \text{m s}^{-1}$
9. (a) $0.628 \, \text{m s}^{-1}$
    (b) 15.9 cm; rope is of negligible thickness.

Exercise 2b – p. 22
1. 51.2   2. 432
3. (i) $\overrightarrow{PO}$ (ii) $\overrightarrow{QO}$ 4. 30
5. 13.9
6. (a) $0.0338 \, \text{m s}^{-2}$ towards the centre of earth
    (b) $0.0169 \, \text{m s}^{-2}$ towards, and at right angles to, the earth's axis
7. $\frac{48\,200}{g}$ m   8. $4.2 \, \text{rad s}^{-1}$ 9. $9.49 \, \text{m s}^{-1}$
10. $8 \, \text{m s}^{-1}$ (below $8.94 \, \text{m s}^{-1}$ for safety)

Exercise 2c – p. 26
1. 10 N   2. 16.7 m
3. $2.19 \, \text{m s}^{-1}$ 4. 112.5 N
5. 8 N
6. (a) 42.7 N, 3.92 N   (b) $14.4 \, \text{rad s}^{-1}$
7. (a) horizontal: 1130 N; vertical: 588 N
    (b) 1280 N at 27° to the horizontal
8. (a) $9.5 \, m$ N   (b) $7860 \, \text{m s}^{-1}$
    (c) $1.21 \times 10^{-3} \, \text{rad s}^{-1}$ (d) 1.44 hours

Exercise 2d – p. 31
1. (a) 288 N   (b) 84°
2. (a) 21.6 N   (b) $3.9 \, \text{rad s}^{-1}$
3. 0.392 m   4. (a) 3 N   (b) $1.73 \, \text{rad s}^{-1}$
5. (a) 9.8 N   (b) 4.9 m   (c) 1.15 m
6. (a) $T = 2 m l \omega^2$, $R = mg$
    (b) $T = 2 m l \omega^2$, $R = m(g - l\omega^2)$
7. (a) (i) $\frac{2}{3}$ (ii) $1.5a \cos \theta$
    (b) towards A: $\frac{15}{4} mg$;   towards B: $\frac{9}{4} mg$
8. $\frac{2}{3} \sqrt{30 ga}$
9. (a) 2.29 N   (b) $1.71 \, \text{m s}^{-1}$
    (c) 32.6 m   (d) 26.4 N
10. (a) 60°   (b) 3.95 N   (c) 5.87 N
11. (a) $\sqrt{gr \tan \theta}$ (b) $94 \, \text{km h}^{-1}$
    (c) Pendulum bob is at a constant distance from side of carriage but string is inclined to the vertical, in a plane parallel to the track, with the bob behind the vertical.

Exercise 2e – p. 36

In each question the object or person is treated as a particle and it is assumed that there is no air resistance.

1. $0.626\,\text{rad}\,\text{s}^{-1}$ 2. $4.43\,\text{rad}\,\text{s}^{-1}$
3. (a) $4800\,\text{N}$
   (b) $32\,\text{m}\,\text{s}^{-1}$; possible overturning ignored
4. (a) $91\,\text{N}$ (b) 1.9 (c) No
5. Horiz: $570\,\text{N}$;   vert: $390\,\text{N}$
6. (a) $mr\omega^2$ (b) mg
   (c) $\sqrt{g/\mu r}$ (d) $3.13\,\text{rad}\,\text{s}^{-1}$; $7.83\,\text{m}\,\text{s}^{-1}$
7. (a) $1.5\omega\,\text{m}\,\text{s}^{-1}$, $3\omega\,\text{m}\,\text{s}^{-1}$, $4.5\omega\,\text{m}\,\text{s}^{-1}$
   (b) (i) $225\omega^2\,\text{N}$
       (ii) $375\omega^2\,\text{N}$
       (iii) $450\omega^2\,\text{N}$
8. (a) $490\,\text{N}$ (b) $37°$ (c) $3.0\,\text{m}$;
   the rope does not stretch and lies in the vertical plane containing the girl's weight and the pole.
9. (a) $58.8\,\text{N}$ (b) $8.85\,\text{m}\,\text{s}^{-1}$

Exercise 2f – p. 42

1. (a) $11\,\text{kN}$ (b) outer
2. (a) outer   (b) $620\,\text{m}$
3. $11\,\text{m}\,\text{s}^{-1}$ $(40\,\text{km}\,\text{h}^{-1})$
4. (a) $\frac{1}{2}$ (c) $22°$
5. $44\,\text{m}\,\text{s}^{-1}$ $(160\,\text{km}\,\text{h}^{-1})$;
   $77\,\text{m}\,\text{s}^{-1}$ $(280\,\text{km}\,\text{h}^{-1})$
6. $8200\,\text{N}$; $0.015\,\text{m}$
7. Between $8.8\,\text{m}\,\text{s}^{-1}$ and $38\,\text{m}\,\text{s}^{-1}$
8. (a) $\dfrac{V^2}{3g}$ (b) $\dfrac{V\sqrt{3}}{3}$
9. $6.6\,\text{m}\,\text{s}^{-1}$; treat motor cyclist and machine as a particle; assume wall to be uniformly rough.
10. (a) When the aircraft is banked, the lift force has a component towards the centre of the circle.
    (b) $49°$. No wind, lift force constant.
11. (b) $\dfrac{V_2^2}{rg} = \dfrac{1-\mu}{1+\mu}$ (c) $r = \dfrac{V_1 V_2}{g}$
    (d) $\sqrt{V_1 V_2}$

CHAPTER 3

Exercise 3a – p. 46

1. (a) $10\,\text{N}:0$, $130\,\text{N}:2400\,\text{J}$,
      $65\,\text{N}: -500\,\text{J}$, $80\,\text{N}: -1600\,\text{J}$
   (b) $300\,\text{J}$ (c) $300\,\text{J}$ (d) $10.8\,\text{m}\,\text{s}^{-1}$
2. $4120\,\text{J}$ 3. $141\,\text{kJ}$
4. (a) $16\,\text{J}$ (b) $36\,\text{J}$
5. (a) $203\,\text{J}$ (b) $33.8\,\text{N}$ (c) 0.0765

6. $18\,\text{m}\,\text{s}^{-1}$ 7. (a) $2840\,\text{N}$
8. (a) $3.76\,\text{s}$ (b) $18.9\,\text{kW}$
9. $4.85\,\text{m}\,\text{s}^{-1}$; assume that Sue is a particle, rope is light, no air resistance
10. (a) $6.26\,\text{m}\,\text{s}^{-1}$ (b) $3.13\,\text{m}\,\text{s}^{-1}$
11. (a) $\mu mgd \cos\theta$
    (b) $mgd \sin\theta$
    (c) $mgd(\mu\cos\theta + \sin\theta) + \frac{1}{2}m(v^2 - u^2)$

Exercise 3b – p. 51

1. (a) $0.533\,\text{J}$ (b) $1.2\,\text{J}$ (c) $2\,\text{J}$
2. (a) $\frac{1}{5}mga$ (b) mga (c) $\frac{5}{4}mga$
3. (a) $2.4\,\text{J}$ (b) $3\,\text{J}$
4. (a) $96\,\text{N}$ (b) $24\,\text{J}$
5. $0.01\pi\,\text{J}$
6. (a) $4.83\,\text{m}$ (b) $33.6\,\text{J}$
7. (a) $56.3\,\text{J}$ (b) $2.81\,\text{W}$
8. (a) $11.8\,\text{N}$ (b) $0.443\,\text{J}$
9. $4\lambda a$
10. (a) $20.4\,\text{N}$ (b) $11.8\,\text{J}$

Exercise 3c – p. 59

1. (a) $4.5\,\text{J}$ (b) $0.125\,\text{J}$
   (c) $4.375\,\text{J}$ (d) $2.09\,\text{m}\,\text{s}^{-1}$
2. (a) 0 (b) $100\,\text{J}$
   (c) $3.92\,\text{J}$ (d) $9.8\,\text{m}\,\text{s}^{-1}$
3. (a) 0 (b) $600\,\text{J}$ (c) $150\,\text{J}$
   (d) $450\,\text{J}$ (e) $0.5\,\text{m}$
4. (a) $100\,\text{J}$ (b) $25\,\text{J}$ (c) $24\,\text{J}$
   (d) $51\,\text{J}$ (e) 0.58
5. (a) (i) $6.4\,\text{J}$ (ii) $7.84\,\text{J}$ (iii) 0
   (b) 0 (c) $3.77\,\text{m}\,\text{s}^{-1}$
6. (a) 0,   (i) 0,   (ii) $10(2 + x)$
   (b) 0; $30x^2$ (c) $3\,\text{m}$
7. (a) $14\,\text{N}$ (b) $3.17\,\text{m}\,\text{s}^{-1}$
8. (a) $6a$
9. (a) $0.49\,\text{m}$ (b) $3.49\,\text{m}\,\text{s}^{-1}$
   (c) $2.59\,\text{m}$
10. (a) $0.3\,\text{J}$ (b) $0.3\,\text{J}$ (c) $2.24\,\text{m}\,\text{s}^{-1}$
    (d) $0.225\,\text{J}$ (e) $23.7\,\text{cm}$
11. $445\,\text{kN}$
12. (a) $1.3\,\text{m}$
    (b) (i) $16\,\text{m}\,\text{s}^{-1}$ (ii) $2.1\,\text{m}$

Exercise 3d – p. 63

1. Assume climber as a particle; no air resistance or wind
   (a) (i) $41\,\text{kN}$ (ii) $31\,\text{m}\,\text{s}^{-1}$
   (b) Rope likely to have a significant weight; air resistance or wind quite likely
2. (a) $d = v\sqrt{Ma/\lambda}$
   (d) When $R = 0$, $d = v\sqrt{Ma/\lambda}$

3. Hooke's Law valid throughout; no wind or other resistance.

(a) $\lambda = 68\,600\,\dfrac{a}{x^2}$ (b) $e = 686\,\dfrac{a}{\lambda}$

(c) $\sqrt{\dfrac{\lambda e}{70a}(2a + e)}$

(d)

a	10	20	30	40	45
x	40	30	20	10	5
λ	429	1520	5150	27 400	123 000
e	16.0	9.03	4.00	1.00	0.028
$2a + e$	36.0	49.0	64.0	81.0	90.0
V	18.8	21.9	25.0	28.2	29.7

(e) Hooke's Law unlikely to apply throughout for shorter ropes.

(f) Allow for weight of rope.

CONSOLIDATION A

Miscellaneous Exercise A – p. 66

1. D   2. A
3. (i) T (ii) F (iii) F (iv) T (v) F
4. $\frac{1}{3}l$
5. (a) $2mg$
7. (a) 11 cm   (b) 12 cm
8. 112.0 m
9. (a) $3Mg$ (b) $48°$
10. (b) $6.64\,\mathrm{m\,s^{-1}}$ (c) 4 m
11. (a) $3mg$ (b) $\frac{7}{3}mg$
12. (a)

$\nearrow T$

$\downarrow 0.25\,g$

(b) 0.392 N   (c) $12.2\,\mathrm{m\,s^{-2}}$
(d) $1.95\,\mathrm{m\,s^{-1}}$ (e) $\omega = \sqrt{g/h}$
(f) If T is horizontal it cannot balance the weight of the particle.
13. (a) 11 760 N (b) 27.4 m (c) $20\,\mathrm{m\,s^{-1}}$; no air resistance; Hooke's Law valid throughout
14. $58°$ (nearest degree); $\tan\theta$ is independent of m and M
15. Air resistance zero
16. (a) $5Mg$ (b) $20Mg$
17. (a) $\frac{1}{2}m(d\omega^2 + 2g/\sqrt{3})$
    (b) $\frac{1}{2}m(d\omega^2 - 2g/\sqrt{3})$
18. 6.49

19. An overestimate; work done by air resistance assists in stopping the descent.
20. (a)

$\downarrow 30g$

(b) 357 N
(d) It is light and inextensible.
(e) 4.19 s
21. $T = \lambda$; EPE $= \lambda a$; $v = \sqrt{\dfrac{2\lambda a}{m}}$
22. 19 kN
23. (i) The point B is in equilibrium.
    (ii) AB $= 0.8\,\mathrm{m}$; $T = 6\,\mathrm{N}$
    (iii) 2.7 J   (iv) $10\,\mathrm{m\,s^{-1}}$
24. D   25. C
26. B   27. B
28. (i) F (ii) T (iii) T (iv) T (v) F
    (vi) F
29. (ii) $T(\cos\alpha - \cos\beta) = mg$
    $\therefore \cos\alpha > \cos\beta, \therefore \alpha < \beta$
    $\therefore$ AP $>$ half length of string
30. 0.66 N; $1.3\,\mathrm{m\,s^{-1}}$
31. (a) $3Mg$ (b) $\sqrt{3gr}$
32. 0.392; $21.4°$; $35.5\,\mathrm{m\,s^{-1}}$
33. No air resistance
    (i) $5.5\,\mathrm{m\,s^{-1}}$ (ii) 1.9 m
34. (a) EPE – work done by friction > 0
    (b) $\frac{1}{3}\sqrt{ga}$
35. (a) 92 kg   (b) $28\,\mathrm{m\,s^{-1}}$

CHAPTER 4
Exercise 4a – p. 82

1. (a) $2.65\,\mathrm{m\,s^{-1}}$ (b) $2.25\,\mathrm{m\,s^{-1}}$
   (c) $1.08\,\mathrm{m\,s^{-1}}$
2. (a) 40.1 N towards O
   (b) 25.4 N towards O
   (c) 4 N away from O
3. (a) 1.4 (b) 1.98 (c) 2.8 (d) 0
4. $113°$
5. (a) 4.87 (b) 5.24
6. (a) 25.5 N, tension   (b) 29.4 N, tension
7. (a) 41.9 N   (b) 14.7 N
8. $27.9\,\mathrm{m\,s^{-2}}$ rad., $9.8\,\mathrm{m\,s^{-2}}$ tang.
9. (a) $204°$; 1.22 cm   (b) $218°$; 1.84 cm
10. (a) $2.63\,\mathrm{m\,s^{-2}}$ rad.; $4.9\,\mathrm{m\,s^{-2}}$ tang.
    (b) $19.6\,\mathrm{m\,s^{-2}}$ rad.; $9.8\,\mathrm{m\,s^{-2}}$ tang.
    (c) $36.6\,\mathrm{m\,s^{-2}}$ rad.; $4.9\,\mathrm{m\,s^{-2}}$ tang.
11. $u > 2\sqrt{ga}$

Exercise 4b – p. 89

1. $4.48 \, \text{m s}^{-1}$
2. (a) $4.18 \, \text{m s}^{-1}$ (b) $8.69 \, \text{N}$
3. (a) $u > \sqrt{3g}$
   (b) oscillation through $180°$
4. $4.81 \, \text{m s}^{-1}$; $5.29 \, \text{N}$
5. $3\sqrt{ga}$
6. (a) $\sqrt{5ga}$ (b) $4mg$
7. (a) $\sqrt{ga/2}$ (b) $\sqrt{7ga/2}$
8. $0.544 \, \text{m}$; $2.31 \, \text{m s}^{-1}$
9. $62.6 \, \text{m s}^{-1}$
10. 3.13
11. (a) $3\frac{1}{3} m$ (b) $3.61 \, \text{m s}^{-1}$ (c) $8.85 \, \text{m s}^{-1}$
12. (a) $\frac{1}{3} m (5v^2 - 3g)$ (b) 2.42
13. (a) $12.8 \, \text{m}$ (b) $9.82 \, \text{m s}^{-1}$

Exercise 4c – p. 96

1. (a) $12 \, \text{m s}^{-1}$
   (c) (i) $1.16 \, \text{N}$, tension;
       (ii) $0.830 \, \text{N}$, thrust
   (d) $165°$
2. (a) $2.8 \, \text{m s}^{-1}$, $29.4 \, \text{N}$
   (b) $1.2 \, \text{m s}^{-1}$ (c) $20.1 \, \text{N}$ (d) $0.073 \, \text{m}$
3. (a) $1600 \, \text{N}$ (b) $770 \, \text{N}$
   (c) $7600 \, \text{N}$ (d) $15.7 \, \text{m s}^{-1}$
4. (a) $J \gg 7.42 \, \text{N s}$ (b) $\frac{3}{4}\sqrt{7g} \, \text{N s} \, (6.2 \, \text{N s})$
5. $\frac{1}{2} l$; $11:5$
6. Assume catcher and flier each to be a
   particle distant $7 \, \text{m}$ from point of
   attachment of ropes.
   (a) $9.5 \, \text{m s}^{-1}$ (b) $4.6 \, \text{m s}^{-1}$
   (c) $1700 \, \text{N}$ (d) $32°$
7. (a) $1.01 \, \text{m s}^{-1}$ (b) $53.6 \, \text{N}$ (c) $505 \, \text{m s}^{-1}$
8. (a) $350 \, \text{N}$, $6.9 \, \text{m s}^{-1}$ (b) $17 \, \text{m s}^{-1}$
   (c) $1700 \, \text{N}$, $29 \, \text{m s}^{-2}$ (d) $18 \, \text{m s}^{-1}$
   (e) $1500 \, \text{N}$, decrease
9. (a) $\frac{4}{3}\pi \, \text{m s}^{-1}$ (b) (i) $\frac{2}{9}\pi^2 \, \text{m s}^{-2}$ (ii) 0
   (c) $R = 60g \cos\theta + \frac{40}{3}\pi^2$; $S = 60g \sin\theta$
   (d) (i) $720 \, \text{N}$ vert. upwards
       (ii) $460 \, \text{N}$ vert. upwards
10. (a) $12 \, \text{m s}^{-2}$ (b) $15 \, \text{m s}^{-2}$
11. (a) $4.0 \, \text{m s}^{-1}$ (b) $2000 \, \text{N s}$
    (c) (i) $6200 \, \text{N}$ (ii) $4900 \, \text{N}$
12. (a) $2.1 \, \text{m s}^{-1}$ (b) $5.6 \, \text{m s}^{-1}$

   (c) $1.05 \, \text{m s}^{-1}$; $1.05\sqrt{3}$

   (d) $0.17 \, \text{m}$ (e)

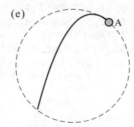

13. (a) $28°$ (b) $1.9 \, \text{m}$ (c) $2.9 \, \text{m s}^{-1}$
    (d) $\longrightarrow 2.58 \, \text{m s}^{-1}$, $1.4 \, \text{m s}^{-1}$
    (e) (i) $6.2 \, \text{m s}^{-1}$ (ii) $0.49 \, \text{s}$ (iii) $1.3 \, \text{m}$

CHAPTER 5
Exercise 5a – p. 101

1. $6 \, \text{N s}$, $\overrightarrow{\text{AB}}$
2. $2 \, \text{m s}^{-1}$, $\overrightarrow{\text{AB}}$
3. $1 \, \text{m s}^{-1}$, $\overrightarrow{\text{BA}}$
4. (a) $12\mathbf{i} - 9\mathbf{j}$ (b) $15 \, \text{m s}^{-1}$
5. $96.3 \, \text{N}$
6. (a) $4 \, \text{m s}^{-1}$ (b) $2.4 \, \text{N s}$ (c) $240 \, \text{N}$
7. Spacecraft $16050 \, \text{m s}^{-1}$,
   rocket $15970 \, \text{m s}^{-1}$ (4 s.f. used in order to
   show the difference)
8. (a) $1.3 \, \text{m s}^{-1}$; $1.8 \, \text{kJ}$ (b) $0.5 \, \text{m s}^{-1}$; $45 \, \text{kJ}$
9. (a) $2.6 \, \text{m s}^{-1}$ downstream
   (b) $0.9 \, \text{m s}^{-1}$ upstream
10. $75\mathbf{i} - 130\mathbf{j}$
11. (a) $0.57 \, \text{m s}^{-1}$ (b) $0.73 \, \text{m s}^{-1}$ (c) $36 \, \text{N s}$
    Father and son treated as particles; no
    friction or air resistance.

Exercise 5b – p. 106

1. (a) $4 \, \text{m s}^{-1}$ (b) $32 \, \text{N s}$
2. (a) $8 \, \text{m s}^{-1}$ (b) $120 \, \text{J}$
3. (a) 0 (b) $14 \, \text{N s}$ (c) $98 \, \text{J}$
4. (a) $12 \, \text{m s}^{-1}$ (b) $48 \, \text{N s}$ (c) 0
5. (a) $2.8 \, \text{m s}^{-1}$ (b) $1.4 \, \text{m s}^{-1}$
   (c) $\frac{1}{2}$ (d) $0.1 \, \text{m}$
6. $2.5 \, \text{m}$; $42 \, \text{N s}$
7. $-6\mathbf{i}$ 8. 0.4 9. $-20\mathbf{i}$
10. $3\mathbf{i} - 5\mathbf{j}$ 11. 0.5
12. (a) $5 \, \text{m}$, $1.25 \, \text{m}$, $0.3125 \, \text{m}$
    (b) $r = 0.25$ (c) $33.125 \, \text{m}$
13. 0.6; ignore friction and mass of puck
14. (a) $\sqrt{ga}$ (b) $\frac{3}{4}\sqrt{ga}$ (c) $\frac{9}{32}a$
15. (a) 0.78 (b) $2.04 \leqslant v \leqslant 2.36$

Exercise 5c – p. 112

1. (a) $u = 7\frac{1}{3}$, $v = 9\frac{1}{3}$; $J = 5\frac{1}{3} \, \text{N s}$
2. (a) 11 (b) $\frac{11}{13}$; $16 \, \text{J}$
3. (a) 2 (b) 3 (c) $25 \, \text{J}$
4. (a) 3 (b) $\frac{2}{3}$
5. $0.2 \, \text{kg}$; $2.5 \, \text{m s}^{-1}$; $0.5 \, \text{N s}$
6. 0.8 7. 0.217% 8. (c) $\frac{1}{2} mu^2 (9 - k^2)$
9. (a) $u = \frac{1}{7}(60 - 10e)$, $v = \frac{1}{7}(4e + 60)$
   (b) $\frac{60}{7} \leqslant v \leqslant \frac{64}{7}$
10. (a) $\dfrac{(k - 3)u}{3(k + 1)}$ (c) $\dfrac{4u}{3(k + 1)}$
    (d) $\dfrac{4kmu}{3(k + 1)}$
11. $\frac{1}{8}(9 - k)$
12. (a) $k = 3$ (b) $2\sqrt{gl}$, $6\sqrt{gl}$ (c) 0
    (d) $6m\sqrt{gl}$

Exercise 5d – p. 118

1. $u_1 = -\frac{4}{3}$, $v_1 = \frac{2}{3}$, $v_2 = \frac{1}{3}$; A travelling away from the wall faster than B so B will not catch A to collide again

2. (a) $u_1 = 3$, $v_1 = 12$
  (b) A: $-3\,\text{m s}^{-1}$, B: $6\,\text{m s}^{-1}$
  (c) B will collide with the wall again.

3. After A strikes B

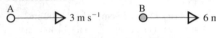

After B strikes wall

After B strikes A

After B strikes wall again

$\frac{1}{2}\text{m s}^{-1}$ ◀━━━━━○ B

There will be no more collisions.

4. After 1st collision

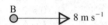 

After 2nd collision

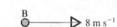

5. 1st:

2nd:

3rd:

6. (a)

(b)

(c) (i) Velocity of B is negative so B hits A.
  (ii) Velocity of B is positive and < velocity of C; no further collisions

7. (a) $\dfrac{m_1}{m_2}u$ (b) $\dfrac{m_1}{m_2}$; (c) $\dfrac{(m_2)^2}{m_1}$

8. $6.12\,\text{m s}^{-1}$ $\left(\frac{300}{49}\right)$

9. $11.4\,\text{m s}^{-1}$ $\left(\frac{343}{30}\right)$

10. (a) $P = 0.25\,(2e - 1)\,\text{m s}^{-1}$, Q: $0.25\,(e + 1)\,\text{m s}^{-1}$
  (b) $e = \frac{4}{5}$; P: $0.15\,\text{m s}^{-1}$, Q: $0.45\,\text{m s}^{-1}$
  (c) P: $0.57\,\text{m s}^{-1}$, Q: $0.09\,\text{m s}^{-1}$

11. (a) $e^2 h$, $e^4 h$, $e^6 h$ (b) GP in which $r = e^2$ (c) $\dfrac{h}{1 - e^2}$

CHAPTER 6

Exercise 6a – p. 122
1. (7.6, 3.2) 2. (6.625, 3.5)
3. 4 cm, 5 cm 4. 4330 km
5. (b) 10 cm from AB, 3 cm from BC
6. $3\frac{1}{3}$ cm from BC, 8 cm from AB
7. (a) $\left(\frac{17}{14}, \frac{19}{14}\right)$ (b) $\left(\frac{26}{5}, \frac{12}{5}\right)$
8. 13.75 cm from AB, 6.25 cm from BC

Exercise 6b – p. 124
1. 22.1 cm
2. 97.4 g
3. 1 g/cm^2

4. (a) $x = 16$ cm;
   cannot be done

   (b) $x = 13.6$ cm;
   can be done

Exercise 6c – p. 132
1. (a) $\frac{8}{3}$ (b) $\frac{3}{2}$ 2. (a) $\frac{16}{3}$ (b) $\frac{12}{5}$
3. $\frac{45}{14}$ 4. $\frac{45}{28}h$
5. $\frac{5}{3}$ 6. $\frac{95}{166}$ (0.572)
7. $\frac{14}{5}$ 8. $\frac{18}{11}$ (1.64)
9. $\frac{1}{2}\left(\frac{e^2+1}{e^2-1}\right)$ 10. $\frac{5}{\ln 6} - 1$ (1.79)
11. $\frac{4}{\ln 5}$ 12. 3
13. $\frac{4}{3}$ 14. $\frac{525}{152}a$ (3.45a)
15. $\frac{27}{40}a$
16. (a) $\dfrac{4a^2x^2}{h^2}$ (b) $\dfrac{4a^2x^2}{h^2}\rho\,\delta x$
   (c) $a^2h\rho$ (d) $\frac{3}{4}h$ from O on Ox

Exercise 6d – p. 140
1. 4.14 cm
2. (a) $\frac{5}{3}$ cm (1.67 cm)
   (b) $\frac{15}{4}$ cm (3.75 cm)
4. $\dfrac{h^2 - 24}{2(h+2)}$ cm
5. 4.7 cm
6. $\frac{9}{13}a$

7. (a) $\frac{65}{24}$ cm (2.71 cm)
   (b) $\frac{115}{32}$ cm (3.59 cm)
8. 30°
9. (a) $\dfrac{3(2h^2 - r^2)}{4(3h + 2r)}$ (b) $r = h\sqrt{2}$
10. $\frac{93}{88}a$ (1.06a)
11. 20
12. (a) 13 cm, 57 cm (b) 22 500 cm^3
    (c) (i) 11 cm   (ii) 14 400 cm^3
        (iii) 9425 cm^3 (iv) 55.2 cm
    (d) (i) B   (ii) B

CHAPTER 7

Exercise 7a – p. 143
1. $P = 2.91$, $\theta = 18°$
2. 34 N at A, 21 N at B
3. 0.532
4. (a) 6.14 kg (b) 0.011 m
5. (a) 53° (b) 233
6. 70 kg; 14 kg extra
7. 4510 N in AC, 6030 N in BC

Exercise 7b – p. 150
1. (a) 72 N (b) 250 N (c) 72 N
   (d) 0.29
2. 53°
3. (a) 9.8 N
   (b) 35.3 N at 76° to
       horizontal

4. $\mu = \frac{9}{8}$; $F = 26\frac{2}{3}$ N
5. (a) 81.3 N
   (b) 49.4 N at 81° to the downward
       vertical.
6. (a) $\frac{1}{2}W$ (b) $\frac{1}{3}\sqrt{3}$ (0.577)
7. (a) 20.4 N
   (b) 40.9 N at 37° to the upward vertical.
8. 67.5

Exercise 7c – p. 154
1. 38°
2. 74°
3. 88°
4. (a) $2\sqrt{6}$ cm (4.90 cm) (b) 8.14 cm

Exercise 7d – p. 159
1. (a) 1.75 cm   (b) Yes
2. (a) 9.11 cm   (b) rest in equilibrium
3. It topples ($\bar{x} = 8.94$ cm which $> AB$)
4. (a) $\dfrac{3 + 8d}{2(3 + 2d)}$ (b) Yes, $\frac{19}{14} > 1$
5. $d \leqslant \frac{3}{4}$
6. (a) $\dfrac{k^2 + 20k + 250}{2k + 50}$
  (b) (i) Yes, $\frac{775}{80} < 10$ (ii) No, $\frac{1050}{90} > 10$
  (c) $k \leqslant \sqrt{250}$ ($\leqslant 15.8$)
7. will remain in equilibrium
8. (a) $\frac{120}{13}$ cm
  (b) will stay in equilibrium

Exercise 7e – p. 165
1. (a) rest in equilibrium   (b) topple
2. will topple
3. will not topple
4. is on the point of toppling
5. will topple
6. will not topple
7. $20\sqrt{3}$ cm (34.6 cm)
8. 4.08 cm,   8.16 cm
9. (a) 29°   (b) 0.56
10. 10°

Exercise 7f – p. 172
1. 51°
2. (b) (i) 5.39 N   (ii) 219 N
3. $\frac{8}{19}$ (0.421)
4. (iii)
5. (ii)
6. (a) $h = 5\sqrt{3}$ cm (8.66 cm)
  (b) it will remain in equilibrium (this is
    true for any angle)
7. $\frac{1}{3} \tan \theta$
8. (a) $\frac{3}{4} W$ (b) $\frac{1}{2} W$ (c) by toppling
9. By sliding when $\tan \theta = \frac{1}{4}$ (for toppling
  $\tan \theta = 2$)
10. (a) $\frac{50}{3} M$ (b) $\frac{127}{33} M$

CONSOLIDATION B

Miscellaneous Exercise B – p. 177
1. $R = 80 - 50 \sin^2\theta \cos \theta$,
  $F = 50 \sin \theta \cos^2\theta$,
  $N = 50 \sin \theta \cos \theta$; 25 N; 58 N
2. (a) 94 N   (b) 0.47   (c) 133 N

3. (a) 49 N
  (b) vertical: 122.5 N; horizontal: 42.4 N
  (c) 0.35
4. B   **5.** C
6. C   **7.** D
8. B
9. (b) $\frac{3}{16} a$
11. (a) $\frac{3}{8} u$
12. (a) 3.20 m s^{-1} (b) 6.81 m s^{-1}
13. (b) $\frac{1}{2} Mg$
14. (a) $v = \sqrt{[ag(2 \cos \theta - 1)]}$
  (b) $mg(3 \cos \theta - 1)$
15. (i) $\frac{1}{30} \pi$ rad s^{-1} (ii) $\frac{1}{3} \pi$ m s^{-1}
  (iii) constant angular velocity so no
    tangential acceleration
  (iv) $40g \cos \theta - R \cos \phi = 10(\frac{1}{3}\pi)^2$
  (v) R = 390N, $\phi = 30°$
16. B   **17.** C   **18.** T
19. T   **20.** F
21. (i) F   (ii) F   (iii) F   (iv) T
22. (a) 0.92
23. (i) 2.8u   (ii) 6mu   (iii) 0.909
24. (b) $\frac{1}{4} u(3e - 1)$ (d) $\frac{27}{16} mu$
25. (i) $\sqrt{2h/g}$ (ii) $2e\sqrt{2h/g}$
26. (i) 14 m s^{-1} (iii) 2.45 N s
  (iv) 4.29 J
  (v) Each upward speed is
    $e \times$ the previous one.
  (vi) 18th
27. (i) 7.0 m s^{-1} (ii) 6.5 m s^{-1}; 2.2 m
  All resistance reduces speed so a greater
  safe height is possible.
28. (b) $v^2 = \frac{7}{12} gr$ (c) $U^2 = \frac{7}{4} gr$
29. (ii) 11.4 m s^{-1} (iii) 8 m s^{-1} (iv) 0.75
30. D   **31.** A   **32.** D
33. (i) F   (ii) T   (iii) T   (iv) F
34. (b) 38°
35. (i) There must be a horizontal
    component to balance that of T.
  (ii) $T \sin 35° = R \sin \alpha°$,
    $T \cos 35° + R \cos \alpha° = 80°$
  (iii) about 457 N
  (iv) about 563 N along the string away
    from A
  (v) about = 563 N
  (vi) $2T \cos 17.5° \approx 1070$ N
36. B: 3100 N; A: 3500 N; 1700 N, 1300 N
37. (b) $2\frac{2}{3}$ cm
38. $k = 0.71$; the normal reaction is vertical
  and passes through C – therefore is
  collinear with the weight.
39. (b) 38°

40. 7.3 cm; $\mu \geqslant \tan 20°$, i.e. $\mu \geqslant 0.36$
41. (b) 9°
42. (i) must be concurrent (ii) $2W\sqrt{5}$
(iii) $\frac{1}{10}W\sqrt{85}$ (iv) $\frac{1}{6}W\sqrt{2}$; 45°
43. Uniform bridge, light rope, smooth pulley
(i) 87 kg (ii) 1400 N

CHAPTER 8
Exercise 8a – p. 193
1. (a) $43\,\mathrm{m\,s^{-1}}$ (b) $7\,\mathrm{m\,s^{-1}}$
(c) 52 m (d) 16 m
2. (a) $v = 30 - 10t$ (b) $s = 30t - 5t^2$
(c) $t = 3$ and height is 45 m
(d) $t = 6$
(e) 2 seconds, from $t = 2$ to $t = 4$
(f) $s = -35$, $v = -40$
(g) 8 seconds
(h) $40\,\mathrm{m\,s^{-1}}$
(i)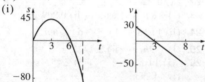

3. (a) $v = 3t^2 - 18t + 24$
(b) $24\,\mathrm{m\,s^{-1}}$
(c) $t = 2$, $s = 20$; $t = 4$, $s = 16$
(d) $a = 6(t - 3)$
(e) (i) $t = 3$ (ii) $v = -3$
(f) $-3\,\mathrm{m\,s^{-1}}$; 0
4. (a) $v = 12 - \dfrac{4}{(1+t)^2}$ (b) $s = 12t + \dfrac{4}{1+t}$
(c) $12\,\mathrm{m\,s^{-1}}$
5. (a) $a = 60 - 12t$
(b) $150\,\mathrm{m\,s^{-1}}$
(c) $450\,\mathrm{m\,s^{-1}}$
(d) $s = 30t^2 - 2t^3 = 2t^2(15 - t)$
(e) $t = 10$, 1000
(f) $t = 15$, $v = -450$, 2000 m
(g)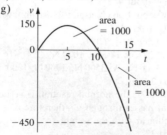

6. (a) $v = 2\mathbf{i} + 3t^2\mathbf{j}$ (b) $\mathbf{a} = 6t\mathbf{j}$ (c) $2\,\mathrm{m\,s^{-1}}$
(d) (i) $\sqrt{73}\,\mathrm{m}$ (ii) $\sqrt{148}\,\mathrm{m\,s^{-1}}$ (iii) 81°
(e) $8y = (x + 1)^3$
7. (a) $\mathbf{v} = 9\mathbf{i} - \frac{16}{t^2}\mathbf{j}$
(b) $\mathbf{r} = 9t\mathbf{i} + \frac{16}{t}\mathbf{j}$
(c) $\frac{4}{3}$
(d) $9\,\mathrm{m\,s^{-1}}$ parallel to $\mathbf{i}$

Exercise 8b – p. 199
1. (a) $v^2 = 2(s^2 + 5s + 2)$
(b) $4\,\mathrm{m\,s^{-1}}$ (c) 1 (or -6)
2. (a) $v^2 = 6s^2 + 8s + 9$
(b) $7\,\mathrm{m\,s^{-1}}$ (c) 1.10
3. (a) $v^2 = s^2 - 8s + 160$ (b) 12.6, $13\,\mathrm{m\,s^{-1}}$
(c) $s = 4$ (d)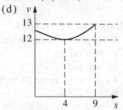

4. (a) $v = 2s^2$ (b) 5
5. (a) $v = 9s^2$ (b) $a = 162s^3$
(c) 10
6. (a) $v^2 = 25s^2 - 576$ (b) 5.2 (c) 4.8
7. (a) $v^2 = 72s - 8s^3 = 8s(9 - s^2)$
(b) $8\,\mathrm{m\,s^{-1}}$ (c) -3, 0, 3
(e) (i) $36\,\mathrm{m\,s^{-2}}$ (ii) $-72\,\mathrm{m\,s^{-1}}$
(f) $\sqrt{3}$; $\sqrt{48\sqrt{3}}\,\mathrm{m\,s^{-1}}$ (9.12 m s^{-1})
(g) Oscillates between $s = 0$ and $s = 3$
8. (a) $v^2 = 2(e^s + 1)$
(b) 10.5 (c) 5.29
9. (a) $v^2 = 144 - 80e^{-s}$
(b) 1.25 (c) 12
10. (a) $v^2 = 20\ln(s + 1) + 16$
(b) 6.94 (c) 601
11. (a) $8\,\mathrm{m\,s^{-1}}$
12. (a) $v^2 = 21 + \dfrac{8 \times 10^5}{x}$
(b) $10.0\,\mathrm{km\,s^{-1}}$
(c) $\sqrt{21}\,\mathrm{km\,s^{-1}}$ (4.58 km s^{-1})

Exercise 8c – p. 204
1. (a) $a = -\dfrac{2}{s^5}$ (b) -64
2. 18
3. $a = \dfrac{-50}{(1 + 2s)^3}$; -0.4

4. (a) $s = \frac{1}{4}t^2$
  (b) (i) 10 seconds
    (ii) $10(\sqrt{2} - 1)$ seconds (4.14)
  (c) 2.5

5. (a) $s = 5e^{-\frac{1}{4}t}$ (b) $5e^{-\frac{1}{2}}$ (3.03) (c) 0

6. (a) $c = -2$, $d = 12$ (c) $s = 6(1 - e^{-2t})$
  (d) 6

7. (a) 0.2 (b) $9.8e^{-0.4s}$

8. (a) $s = 2\sqrt{(t+4)}$ (b) $v = \frac{2}{s}$
  (c) $\frac{1}{4}\sqrt{2}$ (0.354)

9. (b) $t = \dfrac{1}{kp}(1 - e^{-ks})$
  (c) $p = 20$, $k = \frac{1}{3}\ln\frac{1}{2}$ (or $-\frac{1}{3}\ln 2$)
  (d) 6 m

10. (c) $\lambda = 0.08$, $\mu = 0.0283$
  (d) 1st, 7.35s; 2nd, 6.03s
  (e) 1st, 27.5s; 2nd, 20.8s; strengthens
    case for 1st model.

Exercise 8d – p. 208

1. (a) $s = (2t + 6)^{1/3}$ (b) 29
2. (a) $s = 8t + 3e^{-2t} - 3$ (b) 7.19
3. (a) $a = -36\cos 3t$ (b) -2.55
  (c) $s = 4\cos 3t$ (d) 3
4. (a) (i) $v = 5t^2 - 40t$
    (ii) $t = 0$ and $t = 8$
    (iii) $+100\,\text{m s}^{-1}$, $t = 10$;
      $-80\,\text{m s}^{-1}$, $t = 4$
    (iv)

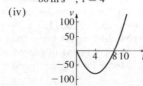

  (b)

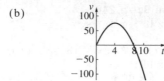

5. (a) $10\frac{2}{3}\,\text{m s}^{-1}$ (b) $13\frac{1}{3}\,\text{m}$
6. (a) $\mathbf{a} = 6t\mathbf{i} - 4\mathbf{j}$ (b) $\mathbf{R} = 8\mathbf{j}$
7. (a) $\mathbf{v} = 2e^{2t}\mathbf{i} + 2t\mathbf{j}$
  (b) $\mathbf{a} = 4e^{2t}\mathbf{i} + 2\mathbf{j}$
  (c) $2\mathbf{i}$; $4\mathbf{i} + 2\mathbf{j}$
  (d) $t = \frac{1}{2}\ln 5$, $\mathbf{v} = 10\mathbf{i} + \ln 5\,\mathbf{j}$
8. (a) $\mathbf{r} = 5\sin 2t\,\mathbf{i} + 5\cos 2t\,\mathbf{j}$
  (b) $\mathbf{a} = -20\sin 2t\,\mathbf{i} - 20\cos 2t\,\mathbf{j}$
  (c) $5\mathbf{i}$ (d) 5 (e) 10 (f) -4
  (g) $x^2 + y^2 = 25$
9. (a) $12.5\,\text{m s}^{-1}$
  (b) $25.8\,\text{m s}^{-1}$, 200 m
  (c) 400 m

10. (b) $s = k\{\frac{1}{4}t^4 - 6t^3 + 36t^2\}$; $k = \frac{1}{4}$
  (c) Yes
  (d) $t = 2.53$, $20.8\,\text{m s}^{-1}$;
    $t = 9.46$, $20.8\,\text{m s}^{-1}$

Exercise 8e – p. 219

1. (a) $\frac{2}{3}\pi\,\text{s}$ · (b) $15\,\text{m s}^{-1}$
  (c) $3\sqrt{21}\,\text{m s}^{-1}$ ($13.7\,\text{m s}^{-1}$)
2. (a) $\frac{8}{5}\pi\,\text{s}$ (b) $6\frac{2}{3}\,\text{m}$
  (c) $\sqrt{39}\,\text{m s}^{-1}$ ($6.24\ \text{m s}^{-1}$)
  (d) $6\frac{1}{4}\,\text{m s}^{-2}$
3. (a) 2.5 m (b) $\pi\,\text{s}$ (c) $5\,\text{m s}^{-1}$
  (d) $10\,\text{m s}^{-2}$
4. (a) 5 m (b) $\frac{4}{5}\sqrt{6}\,\text{m s}^{-1}$, $\frac{4}{25}\,\text{m s}^{-2}$
5. (a) 4 s (b) 0.376 m
  (c) $0.541\,\text{m s}^{-1}$, $0.370\,\text{m s}^{-2}$
6. (a) $-12\sin 3t$ (b) $-36\cos 3t$
  (c) $\ddot{x} = -9x$ (d) $\frac{2}{3}\pi$
7. (a) $x = \frac{1}{2}\cos(\frac{1}{4}\pi t)$
  (b) $v = -\frac{1}{8}\pi\sin(\frac{1}{4}\pi t)$
  (c)

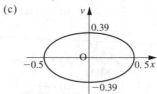

8. 1.33 s
9. $a = 0.023\,\text{m}$, $\ddot{x}_{max} = 3.65\,\text{m s}^{-2}$,
  $v = 0.262\,\text{m s}^{-1}$
10. (a) $\frac{1}{12}\pi\sqrt{3}\,\text{s}$ (b) $\frac{1}{8}\pi\sqrt{3}\,\text{s}$
  (c) $\frac{1}{12}\pi\sqrt{3}\,\text{s}$
11. (a) 1 m (b) $0.274\,\text{m s}^{-2}$
  (c) $0.453\,\text{m s}^{-1}$
12. (a) 38.7 s (b) 9.16 s
13. (a) $0.161\,\text{m s}^{-1}$,
    $259\,\text{m s}^{-2}$
  (b) $0.139\,\text{m s}^{-1}$,
    $129\,\text{m s}^{-2}$

Exercise 8f – p. 222

1. (a) $v^2 = \frac{8}{7}(2lx - 3x^2)$
  (b) 0, $\frac{2}{3}l$ (c) $\frac{1}{3}l$
2. (b) $7.35 \times 10^6\,\text{kJ}$
3. (a) $v = \lambda t - \frac{1}{2}\mu t^2 + k$,
  $s = \frac{1}{2}\lambda t^2 - \frac{1}{6}\mu t^3 + kt$
  (b) Measure the time taken to reach three
    measured distances from the start and
    use $s = f(t)$
  (c) Check further corresponding values of
    s and t

4. (b) $k = 2 \times 60^3$ (432 000)
    (c) (i) 1800 m   (ii) 1920 m
    (d) (i) 15 m s^{-1} (ii) 16.9 m s^{-1}
    (e) (i) 225 m   (ii) 199 m
    (f) Second model correlates very well.
5. (a) 3 m, 750 minutes
    (b) $7\frac{1}{2}$ minutes past 3, 0.025 m/minute
    (c) 3.48 pm   (d) 8.42 pm
6. (a) 2550 minutes
    (b) 2.47×10^{-3} rad/minute
    (c) 423 000 km

CHAPTER 9

Exercise 9a – p. 229
1. (a) $6(5 - e^{3t})$ (b) 24 N
2. (a) $-\dfrac{1}{(t+1)^2}$ (b) $\dfrac{8}{(t+1)^3}$
3. (a) $8 - 2e^{-2t}$ (b) $8t + e^{-2t} + 4$
4. (a) $F = 3s(1 - \frac{1}{s^4})$ (b) $b = 1$
5. (a) $F = \dfrac{120}{v}$ (b) $v^2 = 80t + 36$
    (c) $v^3 = 120s + 344$
    (d) $v = 26$, $s = 143.6$
6. (a) $v = -6 \sin 3t$ (b) $x = 2 \cos 3t$
    (c) $\ddot{x} = -9x$, SHM
7. (a) 50 J   (b) $\frac{125}{3}$ J
8. $11\frac{1}{2}$ m s^{-1}
9. $\frac{1}{20}d^2$; $\frac{1}{20}(20mgd - d^2)$
10. $T = \sqrt{2mu/k}$
11. (a) $8mk$ (b) $(V^2 - 16k)^{1/2}$
12. (a) 38.3 m s^{-1}
    (b) $s = 75 \ln\left(\dfrac{150g}{150g - v^2}\right)$
    (c) 71.1 m

Exercise 9b – p. 234
1. (a) $p = 20q$
    (b) $v = \frac{1}{800}(pt - \frac{1}{2}qt^2)$
    (c) $p - 10q = 600$
    (d) $p = 1200$, $q = 60$
    (e) 11.3 m s^{-1}; agrees very well with the
        value from the graph
    (f) No, because the character of the
        motion changes after 20 seconds.
2. (a) $10v$ N
    (b) $v = 8(1 - e^{-\frac{1}{8}t})$
    (c) 3.72 m s^{-1}
    (d) (i) ≈ 17 s   (ii) ≈ 35 s   (iii) infinite
        time
    (e) $s = 64(\frac{1}{8}t + e^{-\frac{1}{8}t} - 1)$; 10.3 m

(f) Results in (d) are not realistic and the
    result in (e) is not as big as would be
    expected in practice.
3. (a) $J = 1.1mv$
    (b) 20   40   80   100   150   200
        66   132   264   330   495   660
    (c) The correlation is now much closer
        but all the predicted values are still
        lower than those measured.
    (d) From (c) we deduce that the
        coefficient of restitution has been
        underestimated and should be
        increased. Trying $e = 0.14$ gives very
        close correlation.
4. The two values of k given by the second
    model are quite close, at approximately
    4.75×10^6. The two values given by the
    first model differ by too great a margin.
5. (a)   (i) 50 m s^{-1}
        (iii) $t = 50 \ln \dfrac{50}{50 - v} - v$
        (iv) 5.5 s, 16 s
    (b)   (i) 16
        (ii) $t = 25 \ln\left(\dfrac{2500}{2500 - v^2}\right)$
        (iii) 4.4 s, 11 s
    (c) Model 1 does not agree very well,
        giving values that are too large.
        Model 2 has reasonable agreement.
    (d) (i) 1st: 25.2 s   2nd: 16.8 s
        (ii) Model 2 agrees better than
             Model 1 but at this speed even
             Model 2 is not very good.

Exercise 9c – p. 240
1. 6.63×10^{-11} m^3 kg^{-1} s^{-2}
2. 6.0×10^{24} kg
3. (a) 2.7×10^{-6} rad s^{-1} (b) 3.8×10^8 m
4. 3.42×10^8 m
5. (a) 1.75×10^6 m
    (b) 9.6×10^{-4} rad s^{-1}
    (c) 7.4×10^{22} kg
6. (a) 140 N   (b) 1.7 m s^{-2}
    (c) 2.2 m s^{-1}
7. (b) $u^2 - \dfrac{2k}{R}$ (c) $\sqrt{\dfrac{2k}{R}}$
    (d) 1.1×10^4 m s^{-1}
8. (a) 3.7 m s^{-2}
    (b) 4.3×10^{13} m^3 s^{-2}
    (c) $k = g_1 R^2$ (d) $\dfrac{2g_1 R^2}{(2g_1 R - u^2)}$
    (e) (i) $\sqrt{2g_1 R}$ (ii) 5000 m s^{-1}

Exercise 9d – p. 244
1. (a) $9.79\,\mathrm{m\,s^{-2}}$ (b) $1.70\,\mathrm{m\,s^{-2}}$
2. (a) $0.635\,\mathrm{m}$, $1.43\,\mathrm{m}$ (b) $4.8\,\mathrm{s}$
3. $1.5\,\mathrm{m}$ 4. $0.0018\,\mathrm{m}$
5. 9.803
6. 431 seconds in 24 hours in both cases

Exercise 9e – p. 252
1. (a) $72\,\mathrm{N}$ (b) $90\,\mathrm{N}$
2. (a) $\frac{2}{15}\pi$ seconds   (b) $0.4\,\mathrm{m}$
3. (a) $\frac{1}{5}\pi$ seconds   (b) $0.3\,\mathrm{m}$ (c) $6\,\mathrm{N}$
4. (a) $\frac{1}{3}\pi$ seconds   (b) $5\,\mathrm{m}$
5. (a) $3550\,\mathrm{N}$ (b) $18.8\,\mathrm{m\,s^{-1}}$
6. (a) No $\left(ma\left(\dfrac{2\pi}{T}\right)^2 < \mu mg\right)$

   (b) Yes $\left(ma\left(\dfrac{2\pi}{T}\right)^2 > \mu mg\right)$
7. $2.59\,\mathrm{s}$
8. (a) $R - mg = m\ddot{x}$
   (b) $R = m(g - \omega^2 x)$
   (c) (i) $m(g - a\omega^2)$ (ii) $m(g + a\omega^2)$
   (d) $\dfrac{g}{\omega^2}$
9. (a) 2π seconds   (b) $880\,\mathrm{N}$
10. (a) $\frac{1}{5}\pi$ seconds   (b) $1\,\mathrm{m}$ (c) Yes
11. (b) $1.6\,\mathrm{m}$ (c) $\frac{1}{3}\pi$ seconds
   (d) (i) $1.34\,\mathrm{m\,s^{-1}}$ (ii) $1.8\,\mathrm{m\,s^{-1}}$
12. (a) (i) $0.7\,\mathrm{m}$ and $0.3\,\mathrm{m}$; complete
       oscillations if the spring obeys
       Hooke's Law in this range
      (ii) $1.1\,\mathrm{m}$ only; incomplete oscillations
       as the position when $AP = -0.1\,\mathrm{m}$
       is impossible to reach
   (b) $\frac{2}{5}\pi$ seconds, $0.2\,\mathrm{m}$
13. (a) $AO = 3l$, $OB = 2l$
   (b) (i) $2l - x$, $l + x$
      (ii) $\ddot{x} = -\frac{3g}{l}x$
     (iii) $2\pi\sqrt{\frac{l}{3g}}$, l
14. (a) $3.4\,\mathrm{m}$
   (b) (i) $\ddot{x} = -7x$ (ii) $\frac{2}{7}\pi\sqrt{7}$ seconds
      (iii) Incomplete SHM oscillations for
       $1.4 \leqslant x \leqslant 1.8$, Vertical motion
       under gravity for $AP < 2$
15. (a) $2mg$ (b) $3l$
   (c) $2\pi\sqrt{\frac{l}{g}}$, l
   (d) (i) SHM with smaller amplitude but
       same period
      (ii) Incomplete SHM
16. (a) $23.7\,\mathrm{N}$ (b) $0.105\,\mathrm{kg}$

CONSOLIDATION C

Miscellaneous Exercise C – p. 258
1. C   2. B
3. D   4. C
5. (a) $2\,\mathrm{m\,s^{-1}}$ (b) $(3 - e)\,\mathrm{m}$
6. (a) $2\mathbf{i} - e^{-t}\mathbf{j}$ (b) $4.32\,\mathrm{m\,s^{-1}}$
7. (a)

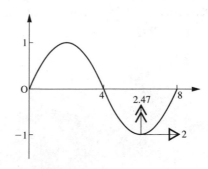

   (b) $\mathbf{v} = 2\mathbf{i}$, $\mathbf{a} = 2.47\mathbf{j}$
8. (a) $-(2\sin 2t)\mathbf{i} - (2\cos 2t)\mathbf{j}$
   (b) $2\,\mathrm{m\,s^{-1}}$
9. (a) $v\frac{dv}{dx} = x^2 + 4$ (c) $55.8\,\mathrm{J}$
10. (a) (ii)

        0.02 —————— v becomes
                 constant

   (b) The initial acceleration is zero so the
       ball descends at constant velocity.
12. (a) $0.3 - 0.6v = 0.03\dfrac{dv}{dt}$
   (d) $0.035\,\mathrm{s}$ (2 sf)
13. (i) $\left(\dfrac{P}{Mk}\right)^{1/3}$ (ii) $\dfrac{P - Mkv^3}{Mv}$
    (iii) $\left[\dfrac{1}{Mk}\left(P - e^{-3kx}\right)\right]^{1/3}$
14. $u^2 = \frac{2}{3}k\ln 3$
15. (a) 6.4×10^{12} (b) $1.46 \times 10^3\,\mathrm{m\,s^{-1}}$
16. (a) $1.62\,\mathrm{m\,s^{-2}}$
17. $v = \sqrt{\left[u^2 + 2k\left(\dfrac{1}{x} - \dfrac{1}{a}\right)\right]}$
18. C   19. D
20. B   21. D
22. C   23. C
24. F   25. F
26. T   27. F
28. T   29. T

30. (a) $2\sqrt{3}\,\mathrm{m\,s^{-1}}$ (b) $1.9\,\mathrm{N}$

31.   (i) $8\,\mathrm{m}$ (ii) $\frac{1}{18}\pi$ seconds
         (iii) $108\sqrt{3}\,\mathrm{W}$

32. (ii) $4\pi\,\mathrm{m\,s^{-1}}$ (iii) $80\pi^2\,\mathrm{m\,s^{-2}}$ (iv) $\frac{2}{3}$

33. (a) $g_1 : g_2 = (3601)^2 : (3599)^2$
       (b) different heights above sea level.

34. (b) $3.8 \times 10^{-3}\,\mathrm{N}$ (c) $0.34\,\mathrm{s}$

35. (a) $\frac{3}{2}L$ (c) $0 \leqslant h \leqslant \frac{1}{2}L$

     (d) $\pi\sqrt{L/2g}$

     (e) $\frac{1}{3}\sqrt{2gL}$

36. (b) (i) 13.22×10^3 seconds
           (ii) $1.41 \times 10^3\,\mathrm{m\,s^{-1}}$
       (c) (ii) $10^4 v = 1.41 \times 10^7\,\mathrm{Ns}$
           (iii) $\theta = 35.2°$

37. $0.28\,\mathrm{V}$

38. (a) (i) Resultant force is zero.
           (ii) Resultant force is perpendicular
               to the tangent on the bends.
       (b) $R = kv$ with $k = 6$
       (d) $10(1 - e^{-t/15})$
       (e) Resistance when just starting unlikely
           to be zero. At high speeds resistance
           quite likely to be almost constant.

INDEX

Acceleration
    as a function of displacement   196, 201
    as a function of time   192
    constant   192
    in circular motion   20, 66
Amplitude of simple harmonic motion   211
Angular velocity   16, 65
    relation to linear velocity   17
Associated circular motion
    for simple harmonic motion   212, 257

Banked tracks   38
    design speed for   39
Bob of a pendulum   39

Centre of mass   121, 176
    calculus methods for finding   126
    of a compound body   134
    of a rigid body   125
    of a semicircle   130
    of a solid cone   127
    of a solid of revolution   128
    of a triangular lamina   121
    standard results   121, 127, 176
Circular motion   20, 78, 84, 175
    force producing   20
    radial acceleration   29, 77
    tangential acceleration   77
    with constant speed   20
    with variable speed   77, 175
Coefficient of restitution   175
Collision   102
Concurrent forces   143
Conditions for equilibrium   6, 143, 145, 176
Conical pendulum   27
Conservation of mechanical energy   45, 52
Conservation of linear momentum   100
Coplanar forces   143

Direct Impact   102
    elastic   102
    inelastic   207
Dot notation   207

Elastic impact   102, 175
Elastic limit   2, 65

Elastic string   1, 65
    modulus of elasticity of   2, 65
    taut   1, 65
    work done in stretching   48
Energy
    conservation of   45
    elastic potential   47, 66
    gravitational potential   45
    kinetic   45
Equilibrium   6, 143, 145
    conditions for   6, 143, 145, 176
    of a body on a horizontal plane   155, 177
    of a body on an inclined plane   161, 177
    of a rigid body   145
    of a suspended body   152, 177
    of concurrent forces   143
    of coplanar forces   145
    of parallel forces   143
    of three forces   143
External impact   104

Gravitation
    Newton's law of   237

Hooke's Law   2, 65

Impact
    direct   102
    elastic   102
    external   104
    inelastic   102, 176
Impacts
    multiple   115
Impulse
    external   104
    instantaneous   100
    of a constant force   100
    of a variable force   225, 258
    relation to momentum   100
Inelastic impact   102, 176
Instantaneous impulse   100

Kinetic energy   45

Lami's Theorem   6

Mathematical models   12, 34, 231
Mechanical energy   45
Modulus of elasticity   2, 65
Momentum   100
    as a vector   100
    conservation of   100
    relation to impulse   100
Motion in a circle   20, 175
    forces producing   20
    radial acceleration   20, 77
    restricted   78
    tangential acceleration   77, 175
    with constant speed   20
    with variable speed   77, 175
Multiple impacts   115

Newton's law of restitution   103
Newton's law of universal
        gravitation   237, 258

Oscillation   213

Parallel forces   143
Particle   12
Perfectly elastic   103, 176
Period of oscillation (SHM)   213
Potential energy
    gravitational   45
    elastic   47, 66
Power   45
Principles
    conservation of mechanical energy   45
    conservation of momentum   100
    impulse–momentum   100
    work–energy   48
Refining mathematical models   232
Restitution
    coefficient of   103
    Newton's Law of   103
Rigid body
    centre of mass of   125
    equilibrium of   145
    suspended in equilibrium   152
Seconds pendulum   243, 257
Sideslip of a vehicle   38

Simple harmonic motion   210, 256
    amplitude   211
    basic equation of   210
    centre   211
    forces causing   245, 257
    oscillation   213
    period of   213
    relation to motion in a circle   212, 257
    standard results   215, 256, 257
Simple pendulum   242, 257
Smooth   12
Spring   3, 65
    compression of   3
    thrust in   3
Solid of revolution   128

Tangential acceleration   77
Tension
    in elastic strings   1
    in springs   3
Terminal velocity   200
Three forces in equilibrium   6
Thrust in a spring   3
Toppling   155, 162
Triangle of forces   6

Universal gravitation
    constant of   237, 258
    law of   237, 258

Variable acceleration   192
    as a function of time   192
    as a function of displacement   196,
        201
Variable force   225, 258
    work done by   225, 258
    impulse exerted by   225, 258
Velocity
    angular   16, 65
    as a function of displacement   201

Work
    relation to mechanical energy   48
    done by a variable force   225, 256
    done in stretching an elastic string or
        spring   48